TABLE OF CONTENTS

Chapter 1: Organizing The Data

Individual values vary from dataset to dataset; no two datasets are the exactly identical. Thus, it is imperative for statisticians and analysts to know how the values of a variable in the dataset are scattered or distributed. How the values of a variable are scattered will in turn determine the shape of the data and the spread of the data points. By knowing the shape and spread of the data, statisticians and analysts can correctly apply the appropriate measure(s) of central tendency or statistical technique to the dataset. We will talk more about that later.

Distribution tells use what are people's opinion about a particular issue - the number or percentage of individuals who strong disagreed, disagreed, were neutral, agreed, or strongly agreed on that issue. Distribution also tells us something about people's health. Generally, are people in good or poor health? We can gage that by looking at the numbers of percentages of individuals that fall under each category (1=poor, 2=fair, 3= good, 4=very good and 5=excellent) of health. Distribution also tells us whether the majority of students (> 50 %) had food poisoning in the past year after eating at the cafeteria. Distribution also tells us something about the socioeconomic well-being of individuals. Just by looking at the variable in a particular dataset, we can see whether the majority of individuals are from low, middle, or upper income groups. If information about the place of residence is collected, we can also indirectly gage the well-being of individuals by looking at the number or percentage of individuals that live in certain types of neighborhoods or certain parts of a county/state/country. The same can be said if information about home values is available in a dataset. As you can see, the distribution of the dataset tells us the ways in which individuals are similar or different from one another. As such, the distribution of a variable tells us something about the disparities and inequalities among individuals in that dataset.

Normal distribution, also known as the Gaussian distribution, is the most common type of distribution for most variables. Even for variables with few categories or limited number of values (e.g. sex, race, age groups and the examples in 2nd paragraph), the distribution of the values of these variables will approximate a normal distribution if we have a large enough sample.

Another common type of distribution would be the binomial distribution. Those who had stats in your undergraduate education may know what *dummy variables* mean. These are essentially variables that only take on two values or variables that have only two categories (i.e. absent or present of an attribute or characteristics). These variables take on the binomial distribution. Binomial distribution takes only two values – whether an attribute, a characteristic, or an experience is present or absent in a person. It is very similar to flipping a coin. If you google the word *dummy*, you will find that it also means "stand-in" or something that's used as a substitute. In statistical sense, especially in quantitative research, dummy is a number that is being used to represent or to substitute for a qualitative characteristic. For instance, we all know that there are two sexes – males and females. We can arbitrarily assign a number for each sex category – one (1) for males and zeroes (0) for females. We can also do it the other way around - one (1) for females and zeroes (0) for males. We can also group students into two categories – those that experienced bullying and those that have not experienced bullying. As in the previous example, we can arbitrarily assign a number for each category – one (1) for yes and zeroes (0) for no.

We can also create a dummy variable from a continuous variable. This is done by dichotomizing the continuous variable into two distinct categories. For instance, we can dichotomize the age variable into minors (under 18) and adults (18 and older). Or we can dichotomize the income variable into below $50,000 and $50,000 and above. Dichotomizing a continuous variable may lead to unrealistic cut-off points, thereby increasing measurement error due to misclassification. The loss of information and reduced statistical power may also result. This is especially so if the relationship between the predictor and the response variable is linear. Therefore, should look at the distribution of the variable before deciding on the cut-off. We have to carefully decide about the cut-off to split the distribution of that variable.

See also p. 82 of *Elementary Statistics in Social Research* on unimodal and bimodal distribution. The latter correspond to binomial distribution. We may also have a multimodal distribution if there are more than two values that stand out in a dataset.

Chapter 2: Measures of central tendency

Means

Unlike the median, a mean (or an average) does NOT take the distribution of the data points into account. We'll talk more about that in the subsequent section.

The presence of extreme data points (we call that *outliers* in statistics) can distort the means. For instance, if we add Bill Gates or Steve Job's income to the existing income data, we are going to blow up the average or mean income data, right?

Those who had stats in your undergraduate education may know what *dummy variables* mean. These are essentially variables that only take on two values or variables that have only two categories (i.e. absent or present of an attribute or characteristics). If you google the word *dummy*, you will find that it also means "stand-in" or something that's used as a substitute. In statistical sense, especially in quantitative research, dummy is a number that is being used to represent or to substitute for a qualitative characteristic. For instance, we all know that there are two sexes – males and females. We can arbitrarily assign a number for each sex category – one (1) for males and zeroes (0) for females. We can also do it the other way around - one (1) for females and zeroes (0) for males. We can also group students into two categories – those that experienced bullying and those that have not experienced bullying. As in the previous example, we can arbitrarily assign a number for each category – one (1) for yes and zeroes (0) for no.

The mean of a dummy variable is the percentage or proportion of individuals having a certain characteristic. In most instances, means are the proportion of individuals where an attribute applies to them or they have experienced something in the past (e.g. bullying, food poisoning, something stolen, etc.). In the second example, the mean of bullying would be the percentage of students in a school or class that experienced bullying in the past. That depends on whether you collected data from every single student in a school or from a class. Let's say that percentage is 67.32 percent in 2014. Thus, we can conclude that on average, 67.32 percent of students in Mr. Thomas's class reported that they experienced bullying in 2014. Note that for dummy variables, the values for the means are always between 0 and 100. Therefore, we can also say that for dummy variables, means can also be the probability that we will observe the person with an attribute, a particular characteristic, or a particular experience in our dataset.

Nevertheless, the interpretation of the mean does NOT always make intuitive sense for all kinds of dummy variables. Let's use the first example again. Let's say that 65 percent of the students in Mr. Thomas's class are females. It does NOT make intuitive sense to say or to conclude that the average student in Mr. Thomas's class is 65 percent female and 35 percent male.

Median

The median, a mean (or an average) takes the distribution of the data points into account. Medians are essentially the 50th percentile. The median appears best to represent the center of the data. It is essentially the midpoint value that separates the data points into two halves. Median is more meaningful than means and is more stable than means as a measure of central tendency. This is mainly because medians are NOT affected by extreme values in the dataset. It makes more intuitive sense to use the median if Bill Gates and Steve Job or any billionaires' incomes are included in a dataset. If we use the mean, it'd definitely blow up the average income, right? If we use the median, we just have to take the midpoint and report on it. Medians can also be when we are unsure about how the data is distributed (non-parametric type of data). Thus, medians can be used for both parametric and non-parametric data.

Mode

A mode is the value that appears most frequently in a dataset. The mode makes intuitive sense when the variable has few categories (e.g. sex, race, age groups). The mode also makes intuitive sense when the variable takes a limited number of values. This often happens in Likert scale variables that asks people whether they strongly disagree, disagree, neutral, agree, strongly agree on a particular stance, in variables that asks people to rate their health (1=poor, 2=fair, 3= good, 4=very good and 5=excellent), or in rating types of questions (1= far from satisfactory, 2 = not satisfactory, 3 = somewhat satisfactory, 4 = satisfactory). With variables that have few categories or take a limited number of values, the analyst can just calculate the number (*frequency*) or percentage of people that fall under each category. It may NOT make intuitive sense to calculate the mean or median for these variables.

When to use each of the measure of central tendency?

Mean and medians are appropriate measures of central tendency when the variable has very specific and unique values like weight, height, income, age, the length of hospital stay, temperature, and so on. In this case, it is almost impossible to use the mode on variables like these because almost every individual has a unique value.

Means can be used when

- there are no individuals with extreme values
- the data is not skewed
- the data is normally distributed

Medians can be used when

- individuals with extreme values (extremely rich/poor, tall/short, thin/obese) are present in the dataset
- the data is skewed
- the data is not normally distributed (or the distribution of the data is uncertain)

Chapter 3: Measures of Variability

Statistics is always concerned about why and how something or someone is different. The variance and standard deviation measure the extent of differences in a given variable. The degree of inequality for a given variable is measured by the standard deviation relative to its mean. Large standard deviations suggest greater inequality or disparity for a given measure or variable. Small standard deviations suggest that the degree of inequality or disparity for a given measure is slighter, minimal, or negligent. Standard deviations of zero indicate that all individuals have the same value for a given variable. For example, if the entire 8th grade class weighs 110 pounds, it would be impossible to test the impact of weight on height because there would be no frame of reference as data for other weights are not available. Hence, we cannot draw conclusions about the impact of weight on height and the statistical programs would give us an error message if we try to do so.

As measures of inequality, standard deviation and variance paints the best and worst picture of a given society. Sampling is done because it is impossible for you to collect data from every individual in the population you are studying. Including everyone in the population in your data collection may take too long and can be too costly. Therefore, standard deviation and variance paints the best and worst of the data we collected a sample of respondents mirrors the population. After doing your best to ensure that the sample corresponds to the larger population on the characteristics of interest like race, gender, age, and marital status, you can apply the conclusions from the sample to the entire population. In other words, you can generalize the findings you obtained from the sample to the larger population.

If you square the standard deviation you get the variance. Or put it differently, if you take the square root of the variance you get the standard deviation.

$$\text{(Standard Deviation)}^2 = \text{Variance} = \sum_{i=1}^{n} \frac{(x - \bar{x})^2}{n-1}$$

When you multiply the standard deviation by itself, it intensifies large values. $1^2 = 1$, $2^2 = 4$, $3^2 = 9$, $4^2 = 16$, and so on. As you can see, the value gets larger as you progress from 1 to 2 and then to 3 and so on.

$$\sqrt{\text{variance}} = \text{Standard Deviation}$$

At times, the standard deviations make more intuitive sense. Thus, the purpose of using the standard deviation is to keep values close to the mean.

The numerator, $(x - \bar{x})$, indicates how far the data point is from the center

Variance makes more intuitive sense than the standard deviation in other instances. Taking the sum of the squared deviations give us a 0. See p. 109, 111, and 404 (3rd and 4th columns). As such, variance is often used in the measures of sums of squares. We square it because we want the negative and positive dispersions to have a role in the measurement scheme.

Why we need to divide n-1 to get an unbiased variance? Why we need to divide n-1 to get an unbiased variance?

$$\sum_{i=1}^{n} \frac{(x - \bar{x})^2}{n-1}$$

One intuitive explanation is that the n−1 in lieu of n in the denominator is a degrees of freedom adjustment stemming from the fact that we are using the sample mean rather than the population mean in our calculation of the sample variance. Using n - 1 is useful for scale measurement.

To prove that you need to have n-1 in the denominator requires basic familiarity with the "expectation operator". You add and subtract the true population mean inside the brackets of the squared term, re-arrange then expand the square before bringing through the E operator.

Chapter 4: Probability and P-Values

Think about it . . . can we be 100% sure about the likelihood that something will occur in life? (I think you know the answer)

We often take calculated risks in life. The purpose of statistics is to try to measure risks or to tame uncertainties by quantify them. This is done by making educated guesses and deciding beforehand how large a risk we are willing to take if we are wrong. Think about it. We often make educated guesses or try to figure out the odds in coin flip, gambling, dice rolling, and so forth. We often make these educated guesses by setting a probability and confidence interval (we will discuss them later) we are comfortable with to estimate the percentages of time we are wrong. These probabilities are called p-values. They are closer to reality than random guesses because they are based on some kind of mathematical logic. A probability is essentially numerical measures of the likelihood that a particular event will occur. We often set a threshold for the (calculated) risks are we willing to undertake. This threshold is called the level of significance of the statistical test, signified by the alpha (α). 0.10, 0.05, and 0.01 are the common thresholds for the level of significance. In most instances, people settle for a 5% risk (α 0.05) and hence 0.05 is the most commonly used cutoff for statistical significance. See p. 232 of ***Elementary Statistics in Social Research***.

Think back to your M & M's example. It is entirely possible that we get several odd samples where the number of M & M's in these bags differs somewhat from what we normally get in a typical M & M's bag. Thus, the means and standard deviation from these samples will vary somewhat from the more typical samples of M & M's. In this instance, the probability is essentially numerical measures of the likelihood that a particular M & M sample will look odd (i.e. too few or too many candies). This probability is still measured by the p-value.

Many people make the mistake thinking that small p-values imply large difference between groups. They only imply that we are pretty confident that there is a difference between two groups or there is an association between two variables. Bear in mind that statistical significance does not always translate into predictive accuracy because the p-value is affected by the sample size and variability. P-values do not tell us the magnitude of the differences. It only tells us the likelihood that the group differences may be due to sampling error instead of being the reflection of the groups coming from different parent populations.

The P-value is often incorrectly interpreted as the probability that the null hypothesis is true. For example, suppose that a study on gender differences in mathematics achievement produced a p-value of 0.0232. This does NOT mean that there is a 2.32% chance that the null hypothesis is true or there is a 2.32% chance of committing a Type I error. The correct interpretation of the p-value would be that there is a probability of 0.0232 that you will incorrectly reject the null hypothesis and claim that there are no gender differences in mathematics achievement. In other words, p-values tell you how likely are your conclusions, under the conditions of the null hypothesis. It does not tell you anything about the research hypothesis. It does not indicate whether the research hypothesis is true. Here, you can see that probabilities are also the rules that link the sample to the population.

The acceptable risk (probability) varies according to the research question at hand. What if the question was about the effectiveness of a new drug? You might want to be more certain before accepting the risk of being wrong. In other instances, there's little risk from being wrong and a 5% risk may be too strict.

Chapter 5: The Normal Curve

The normal curve (*aka* the *bell curve*) is a theoretical or ideal model (see p. 146 of your text). It was developed by Abraham de Moivre. One of his distinguished works revealed that the mortality rate over a person's age formed a bell shape curve (i.e. a normal distribution). The normal curve become one of the most prominent models of probability distributions when his successors found that other common personality and psychological traits (weight, height, income, math / science / reading abilities, years of schooling, satisfaction, measures of abilities, anxieties, depression, scores for introversion/extroversion, etc.) formed a bell shape too. We all know that income, weight, heights, or even personality traits vary from person to person but the overall distribution of income, weight, heights, or even personality traits in any state or country is quite normal. There are only a few individuals with extreme values, right? *You don't have to know or memorize the formula to understand normal curve.*

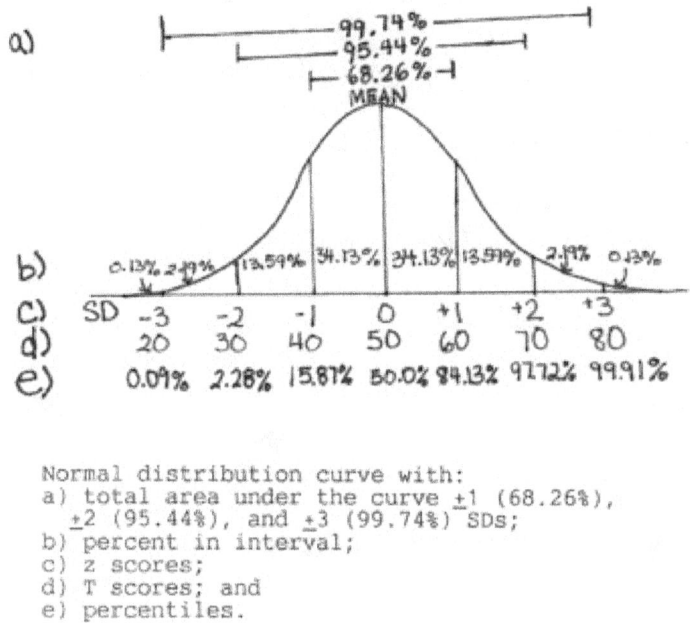

Normal distribution curve with:
a) total area under the curve ±1 (68.26%),
   ±2 (95.44%), and ±3 (99.74%) SDs;
b) percent in interval;
c) z scores;
d) T scores; and
e) percentiles.

Source: http://web.cortland.edu/andersmd/STATS/normcurv.html

As you can see, the normal curve is always symmetry around the mean (μ). Its highest point is also at the mean. It is also bell-shaped where the width is related to the size of the standard deviation. The raw scores (d) are transformed to standardized scores called the z-scores (c) to indicate their position relative to the center (i.e. mean). Because of such transformation, the normal distribution has a mean of 0 and a standard deviation of 1. It's associated probabilities are provided in (e). See the formula on p. 156. Because z-scores are expresses as *the number of standard deviation units from the mean*, the unit of measurement no longer matter.

_____ of standard deviation units above or below the mean (see p. 156) do not give any intuitive sense. Because of that, the obtained z-score needs to be converted to its associated probability to find out the percentages of individuals with raw scores greater or lesser than the raw score. We do that by converting the z-scores to the percentage of area under the normal curve (see Table A on p. 545). Nowadays, we can just consult Dr. Google by typing "drawing a normal curve online" and select from the list of websites. One good one I often use is:

http://homepage.stat.uiowa.edu/~mbognar/applets/normal.html

Here's an example:

Suppose that SAT scores among U.S. college students are normally distributed with a mean of 500 and a standard deviation of 100. **What is the probability that a randomly selected individual from this population has an SAT score at least 600?**

$$z = \frac{x - \mu}{\sigma} = \frac{600 - 500}{100} = 1$$ standard deviation unit above the mean, therefore P (X >= 600) = 0.1587. 15.87% of students have scores greater than 600. This is indicated by the area shaded in pink.

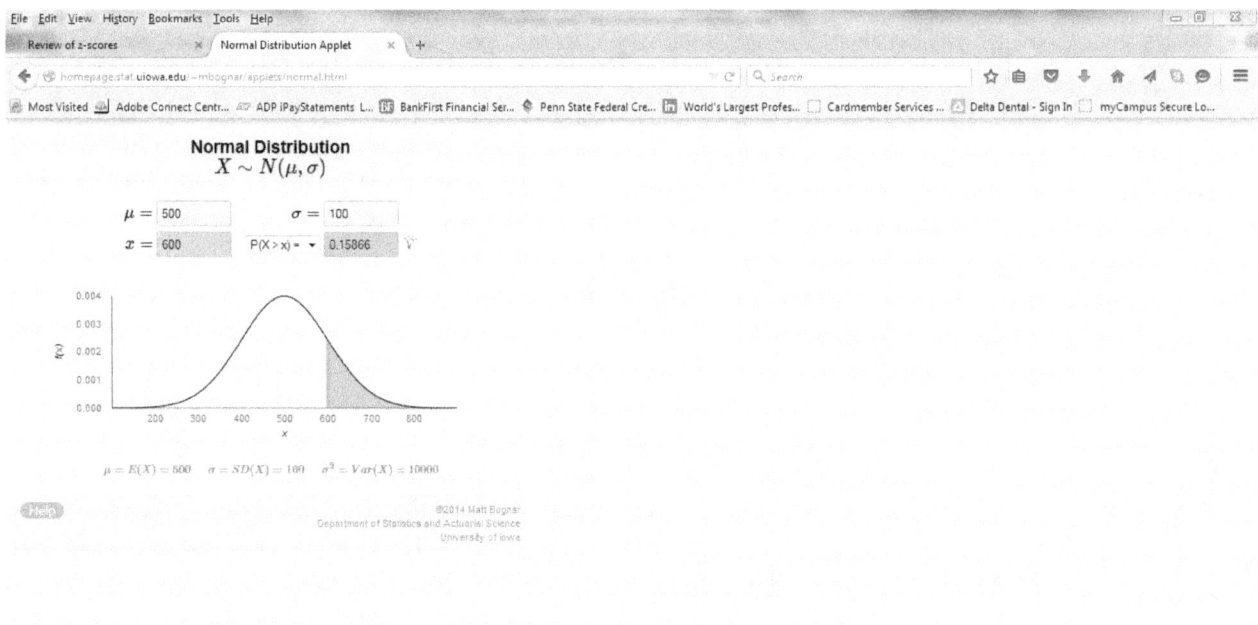

What is the probability that a randomly selected individual from this population has an SAT score at or below 600?

This is indicated by the area shaded in white. The whole normal curve is 100%. Therefore, 100% - 15.87% is 84.13%. We can say that 84.13% of students have scores greater at or below 600.

What is the probability that a randomly selected individual from this population has an SAT score below 450?

$z = \dfrac{x - \mu}{\sigma} = \dfrac{450 - 500}{100} = $ -0.5 standard deviation units below the mean, therefore P (X < 450) = 0.3085. 30.85% of students have scores below 450.

What is the probability that a randomly selected individual from this population has an SAT score between 400 and 700?

We need to do this is 2 steps.

First, we calculate the probability that a randomly selected individual from this population has an SAT score below 400. The probability is 0.15866 or 15.87% of students have scores below 400.

$z = \dfrac{x - \mu}{\sigma} = \dfrac{400 - 500}{100} = $ -1 standard deviation unit below the mean, therefore P (X < 400) = 0.1587. 15.87% of have scores below 400.

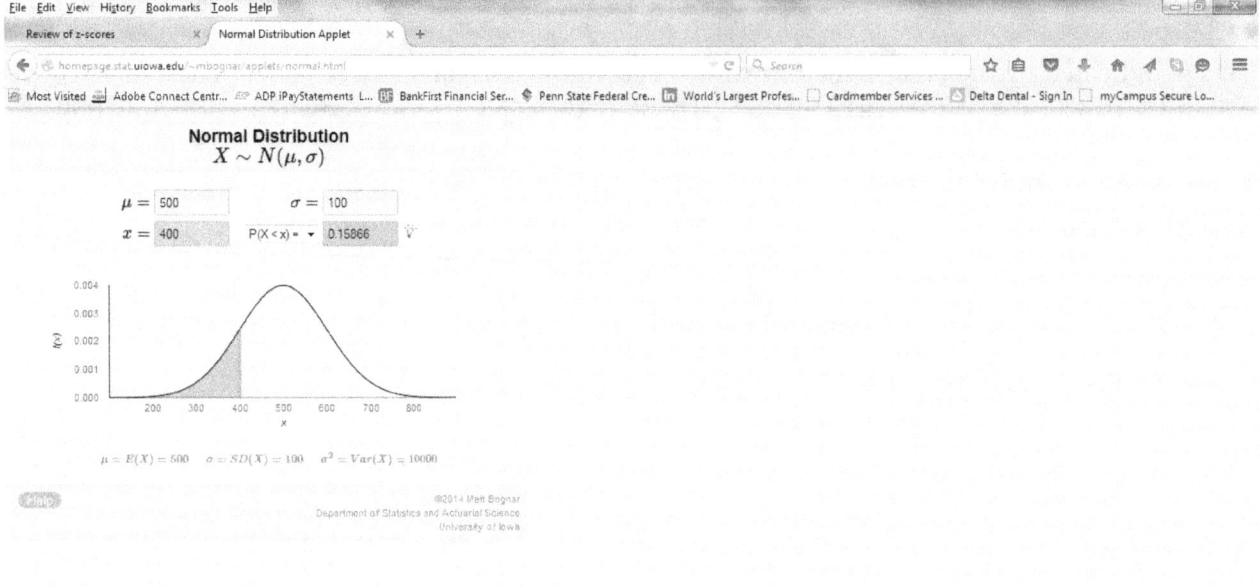

Then, we calculate the probability that a randomly selected individual from this population has an SAT score above 700. The probability is 0.02275 or 2.28% of students have scores above 700.

$$z = \frac{x - \mu}{\sigma} = \frac{700 - 500}{100} = 2 \text{ standard deviation units above the mean, therefore P } (X > 700) =$$

0.0228. 2.28% have scores greater than 700.

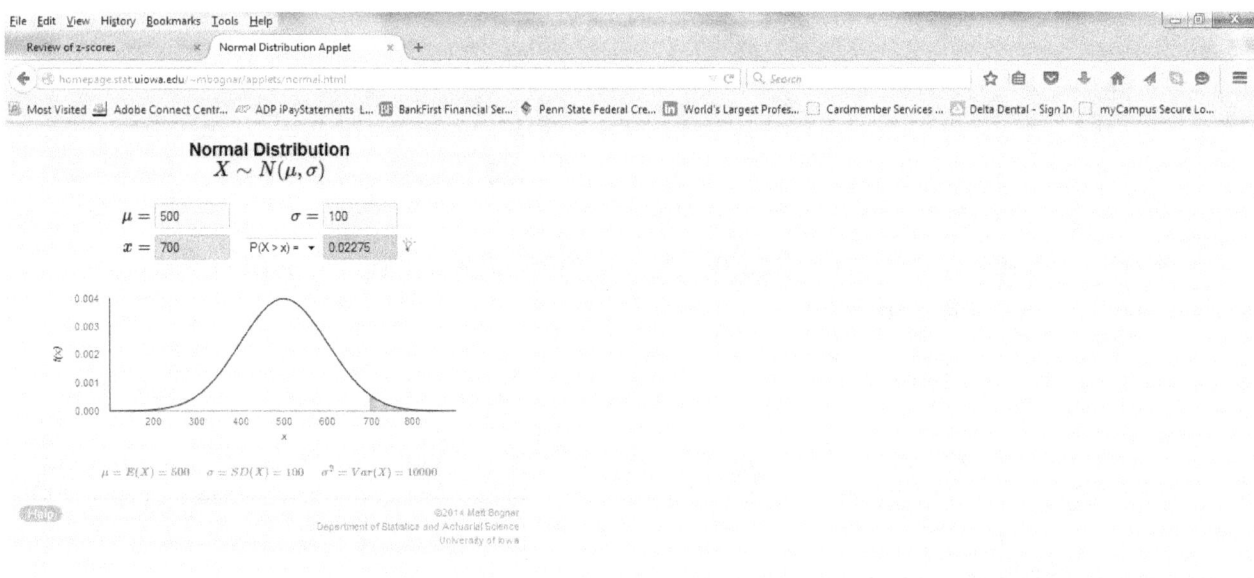

Finally, since the whole normal curve is 100%, we need to subtract 100 from these two percentages, we get the percentage of students who scored between 400 and 700 in their SATs.

100% - 15.87% - 2.28% = 81.85%. Therefore, 81.85% of students who scored between 400 and 700 in their SATs.

This is what we called percentile, the value below a given percentage. The normal curve is also a probability distribution (see p. 146-164 of your text). As I mentioned earlier, z-scores by itself does NOT make intuitive sense. Now you can see how the z-score is just a go-between for a raw score (SAT scores in this example) to its associated percentile.

We talked about binomial distribution last week. Here's how binomial distribution eventually approximates a normal distribution. My Gen Ed course (SOC 101) has about 40 students each semester. Let's have each student toss a coin 10 times and count the number of heads. Draw a line on the blackboard with numbers below to represent the number of heads.

```
 0  1  2  3  4  5  6  7  8  9  10
```

Have the student each report how many heads out of the 10 flips they had. Record them out.

```
                  X
              X   X
              X   X   X
          X   X   X   X
          X   X   X   X   X
      X   X   X   X   X   X   X
  X   X   X   X   X   X   X   X   X
  0   1   2   3   4   5   6   7   8   9   10
```

You will notice that most students report having 4 to 6 heads out of 10 tosses. The diagram approximates a normal (bell-shaped) curve.

Another example would be the demand for a specific type of ice-cream at a store:

y	0	1	2	3	4	5	6	7
P(y)	0.1	0.2	0.3	0.15	0.1	0.05	0.05	0.05

Where y = the number of people who demanded for that ice-cream on a given day and p(y) = the probability associated with the number of demands for a given day.

You see that the probability for demanding that specific type of ice-cream peak between 2-4 times a day.

A good explanation of the normal curve can be found at:

https://statistics.laerd.com/statistical-guides/standard-score.phphttps://www.mathsisfun.com/data/standard-normal-distribution.html

The history of the normal curve can be found at:

http://www.stat.wvu.edu/srs/modules/normal/normal.html

Chapter 6: Confidence Intervals

To ascertain how precise is the estimate of effects (for instance, the mean and the regression coefficient), a confidence interval (CI) is formulated. Confidence intervals are created to show the likely range of an estimate. The estimate can be a mean, a proportion, a percentage, or any parameter that describes or summarizes the data. Confidence intervals are also used to show the likely range for the size of an effect or for the strength of a relationship. Confidence intervals show the extent to which an estimate could be accurate. They can be used for any type of statistic - means, medians, correlations, percentages, regression coefficients, and so on. The CI is constructed to establish a range of values for the unknown estimated population parameter, as well as the probability of being right (the degree of confidence for this estimate). The 95% or 99% CI is most commonly used.

The larger the sample size, other things remaining equal, the narrower/smaller the confidence interval. The 95% confidence interval is wider than the 90% confidence interval. Narrower confidence intervals are more precise. The 95% confidence interval has a larger margin of error[1] than the 90% confidence interval, which means that the level of precision is lower. Thus the 90% confidence interval has a higher level of precision because it has a smaller margin of error than the 95% confidence interval.

The Relationship Between Confidence Intervals And Significance Tests

The confidence interval is constructed to show the extent to which an estimate could be accurate by providing the likely range of an estimate. The standard practice is to construct confidence intervals at the 95% level – so this means that we are 95% confident (sure) that the true effect size lies within the boundaries specified by the confidence interval. This corresponds to p-values which estimate the percentages of time we are wrong. 0.05 is often regarded as the most commonly used cutoff for statistical significance.

In terms of significance testing, confidence intervals may be better than p-values because it consists of the point estimate (the most likely estimate of the population parameter) and a margin of error around that point estimate. As such, confidence intervals can provide good estimates of the unknown population parameter because the intervals constructed tend to contain the parameter.

---

[1] The range of values above and below the sample statistic or the amount of uncertainty that surrounds the sample estimate of the population parameter.

Chapter 7: t-test and ANOVA

t-tests and analysis of variance (ANOVA) are commonly used to compare means for two or more groups. Since both are parametric statistical techniques, the dependent variable should be measured on continuous interval or ratio level. In both t-test and ANOVA, the dependent variable is always an interval or ratio level continuous and the independent variable is always categorical with two or more different levels/categories. The levels correspond to different groups.

The t-test compares the means between two groups. Those two groups can be either two different groups of individuals (e.g. single and married, male and female, smoker and non-smoker, obese and normal weight, etc.) or data collected on the same individuals at two time points (e.g. 2000 and 2015, January 2015 and April 2015, etc.). The former constitute independent samples and the latter constitute dependent samples.

Independent samples

Independent t-test is used for independent samples. It assesses whether or not two groups of people have different mean values on the same variable. That variable has to be a continuous variable. For instance, independent t-test is used to compare the mean mathematics achievement scores between male and female students. Independent t-tests can also be used to compare the mean depression scores of smoker and non-smoker. We know that smoking is related to depression, right? Still, independent t-tests can be used to assess how obese and normal weight students differ in their mean test scores (like those obtained from a math or reading test). Note that we use independent t-test in these examples because male and female students, single and married individuals, smokers and nonsmokers and so forth are from two different groups. A person cannot belong to two groups at the same time. The formula for independent t-test is:

$$t = \frac{\overline{X}_1 - \overline{X}_2}{S_{\overline{X}_1 - \overline{X}_2}}$$ (see p. 238 of ***Elementary Statistics in Social Research***)

Also pay particular attention to how the difference between the two standard deviations, $S_{\overline{X}_1 - \overline{X}_2}$, are obtained on p. 238 of ***Elementary Statistics in Social Research***.

Note that the Levene test is used to assess whether the variances are equal. See p. 242 of ***Elementary Statistics in Social Research***. If variances are unequal, a different formula should be used to assess the difference between the two standard deviations. The formula is provided on p. 242 of ***Elementary Statistics in Social Research***. After that, you proceed to calculate the independent t-test using the same formula (see the bottom of p. 243 of ***Elementary Statistics in Social Research***).

The authors in *Evaluation of the Children at Risk Program.pdf* may have used independent t-tests when comparing the number of extra-curricular activities, antisocial behaviors, violent crimes committed, and drugs sold between adolescents (11 to 13 of age) who participated in the Children at Risk (CAR) program and their counterparts who did not participate in the CAR program.

Independent t-test examples

Levene test is used to assess the equality of variances.

Example 1: Equal variance

Group Statistics					
	Sex	N	Mean	Std. Deviation	Std. Error Mean
Feelings. CESD (Center for Epidemiologic Studies Depression Scale) Sum of 20 items on scale	Female	1068	13.3859	10.42974	.31914
	Male	527	12.3340	9.89423	.43100

Step 1: $\overline{X}_1 = 13.3859$; $\overline{X}_2 = 12.3340$;

Step 2: $s_1^2 = (10.42974)^2 = 108.7795$; $s_2^2 = (9.89423)^2 = 97.8958$

Step 3:

$$s_{\overline{X}_1 - \overline{X}_2} = \sqrt{\left(\frac{N_1 s_1^2 + N_2 s_2^2}{N_1 + N_2 - 2}\right)\left(\frac{N_1 + N_2}{N_1 N_2}\right)} = \sqrt{\left(\frac{(1068)(108.7795) + (527)(97.8958)}{1068 + 527 - 2}\right)\left(\frac{1068 + 527}{(1068)(527)}\right)} = 0.5463$$

Step 4: $t = \dfrac{\overline{X}_1 - \overline{X}_2}{s_{\overline{X}_1 - \overline{X}_2}} = \dfrac{13.3859 - 12.3340}{0.5463} = 1.925$

Step 5: df $= N_1 + N_2 - 2 = 1068 + 527 - 2 = 1593$

Step 6: Step 6: Obtained $t = 1.925$; Table $t = 1.96$; $\alpha = 0.05$

Conclusion: Because the obtained t is greater than the Table t, we reject the null hypothesis at $\alpha = 0.05$ level. Therefore, we conclude that there is enough evidence to suggest that the males and females have different depression scores.

Example 2: Unequal variance

Group Statistics					
	Sex	N	Mean	Std. Deviation	Std. Error Mean
LSNS(Lubben Social Network Scale) Total LSNS score from each of the ten individual items. Score range 0-50. Higher LSNS scores reflect greater level of social support. (scores on each item were anchored between 0 and 5 in order to permit equal weighting)	Female	1065	27.9757	8.73202	.26757
	Male	529	24.9620	9.59959	.41737

Step 1: $\overline{X}_1 = 27.9757; \overline{X}_2 = 24.9620;$

Step 2: $s_1^2 = (8.73202)^2 = 76.248$; $s_2^2 = (9.59959)^2 = 92.152$

Step 3: $s_{\overline{X}_1 - \overline{X}_2} = \sqrt{\dfrac{s_1^2}{N_1 - 1} + \dfrac{s_2^2}{N_2 - 1}} = \sqrt{\left(\dfrac{76.248}{1065 - 1}\right) + \left(\dfrac{92.152}{529 - 1}\right)} = 0.496$

Step 4: $t = \dfrac{\overline{X}_1 - \overline{X}_2}{s_{\overline{X}_1 - \overline{X}_2}} = \dfrac{27.9757 - 24.9620}{0.496} = 6.075$

Step 5: df = the smaller of the two samples = 529

Step 6: Step 6: Obtained $t = 6.075$; Table $t = 1.96$; $\alpha = 0.05$

Conclusion: Because the obtained t is greater than the Table t, we reject the null hypothesis at $\alpha = 0.05$ level. Therefore, we conclude that there is enough evidence to suggest that the males and females have different social network scores.

Test of difference between proportions

	Registered to vote	Not registered to vote	Overall
Sample size (N)	$N_1 = 836$	$N_2 = 129$	$N_1 + N_2 = 965$
Support for the legalization of same sex marriage (f)	$f_1 = 448$	$f_2 = 64$	$f_1 + f_2 = 512$

Proportion that provided support for such legalization (P)	$P_1 = \dfrac{448}{836} = 0.535885$	$P_2 = \dfrac{64}{129} = 0.496124$	$P* = 0.53057$

Source: CBS News/New York Times National Poll

$\underline{\text{Step 1}}$: $P* = \dfrac{N_1 P_1 + N_2 P_2}{N_1 + N_2} = \dfrac{(836)(0.535885) + (129)(0.496124)}{836 + 129} = \dfrac{512}{965} = 0.53057$

Note that f_1 is equal to $N_1 * P_1$ and f_2 is equal to $N_2 * P_2$.

$\underline{\text{Step 2}}$: $s_{P_1 - P_2} = \sqrt{P*(1 - P*)\left(\dfrac{N_1 + N_2}{N_1 N_2}\right)} = \sqrt{(0.53057)(0.46943)\left(\dfrac{965}{(836)(129)}\right)} = 0.047209$

$\underline{\text{Step 3}}$: $z = \dfrac{P_1 - P_2}{s_{P_1 - P_2}} = \dfrac{0.535885 - 0.496124}{0.047209} = 0.842241$

$\underline{\text{Step 4}}$: Step 6: Obtained $z = 0.84$; Table $z = 1.96$; $\alpha = 0.05$

Conclusion: Because the obtained z is greater than the Table z, we fail to reject the null hypothesis at $\alpha = 0.05$ level. Therefore, we cannot conclude that there is enough evidence to suggest that the registered and non-registered voters generally have different opinions about the legalization of same sex marriage.

Dependent samples

In contrast, a paired t-test can be used to assess whether students' test scores improved over time (e.g. after three months or half a year) given their participation in the after school remedial program. A paired t-test can also be used to assess whether Mississippians' DUI offenses increased or decreased over time after participating in the Mississippi Alcohol Safety Education Program (MASEP). Still, a paired t-test can be used to assess whether veteran's depression scores improved over time after participating in a treatment program. Note that as for the case of paired t-tests, we are measuring the same person at two different time points, be it a few weeks, a few months, or a few years apart.

Note that a different formula is used to obtain the difference between the two standard deviations:

$SD = \sqrt{\dfrac{\sum D^2}{N} - (\overline{X}1 - \overline{X}2)^2}$ (see p. 245 of *Elementary Statistics in Social Research*)

We use z-scores to test the differences between two proportions. First, we need to find out the combined sample proportions:

$$P* = \frac{N_1 P_1 + N_2 P_2}{N_1 + N_2}$$ (see p. 250 of ***Elementary Statistics in Social Research***)

Then, we proceed to calculate the difference between the two standard deviations:

$$S_{P_1-P_2} = \sqrt{P*(1-P*)\left(\frac{N_1 + N_2}{N_1 N_2}\right)}$$ (see p. 250 of ***Elementary Statistics in Social Research***)

Finally, we proceed to calculate the difference between the two proportions

$$z = \frac{P_1 - P_2}{S_{P_1-P_2}}$$ (see p. 250 of ***Elementary Statistics in Social Research***)

Pay particular attention to p. 249-252 on <u>*Elementary Statistics in Social Research*</u> *on the steps involved in assessing whether there's a difference between two proportions.*

Note that the hypotheses are formulated differently for one and two-tailed test.

Paired t-test example

Step 1: $\overline{X}_1 = 2.821$; $\overline{X}_2 = 2.646$

Step 2: $S_D = \sqrt{\frac{\sum D^2}{N} - (\overline{X}_1 - \overline{X}_2)^2} = \sqrt{\frac{58.7585}{30} - (2.821 - 2.646)^2} = 1.928$

Step 3: $S_{\overline{D}} = \frac{S_D}{\sqrt{N-1}} = \frac{1.928}{\sqrt{30-1}} = 0.358$

Step 4: $t = \frac{\overline{X}_1 - \overline{X}_2}{S_{\overline{D}}} = \frac{2.821 - 2.646}{0.358019} = 0.489$

Step 5: df = N − 1 = 30 − 1 = 29

Step 6: Obtained $t = 0.489$; Table $t = 2.045$; $\alpha = 0.05$

Conclusion: Because the obtained *t* is lesser than the Table *t*, we cannot reject the null hypothesis at α = 0.05 level. Therefore, we conclude there is not enough evidence to suggest that liquor expenditure is higher in year 2 than year 1.

Other uses of the t-test

The Student's t-distribution is used when the z-scores cannot be used or when we lack information about the population. It is used for small sample sizes (< 30). It is also used when the population standard deviation, σ, is unknown.

z-scores is NOT a function of the sample size. Think back to the formula of z-score:

$z = \dfrac{x - \mu}{\sigma}$. Therefore, z-scores cannot be used when the sample size is small. However, the t-distribution is a function of the sample size (*via* the degrees of freedom).

The t-distribution can be regarded as an approximation of the normal distribution. It has an additional parameter, N. Refer to p. 196-199 of ***Elementary Statistics in Social Research***. You don't have to memorize these formulas but note that N is embedded in the calculation of the population and the sample standard deviations. From these pages, you can also see how the t-distribution is only used to for specific alpha (α) values and it depends on the N and the sample standard deviation, *s*. However, as the sample size (n) increases, s gets closer to the population standard deviation, σ, and the t-distribution approximates the normal distribution.

I mentioned that often times, we cannot observe the true intercept and slope(s) in the population (i.e. the β_0 and β_k). The t-tests can also be used to test the significance of the individual regression coefficients to determine if a linear relationship exists between the dependent variable and that particular independent variable. This is done by dividing the estimated coefficient by its standard error. The resulting ratio tells us how many standard-error units the coefficient is away from zero. It tests the null hypothesis that the population regression slope is 0, that is, there is no relationship between that particular independent variables and the dependent variable. If the *t*-statistic reported for the slope coefficient is larger than zero, we can conclude that it is highly unlikely that the sample we have would have come from a data-generating process in which the true slope is zero. Thus we would reject the null hypothesis that the true slope is 0 and conclude that there is a relationship in the population between that particular independent variables and the dependent variable. Nowadays, the statistical software does that for us.

One-Way ANOVA

While the *t*-test is used to compare the means between two groups, the one-way ANOVA can be used to compare means <u>between three or more groups</u>. These groups are unmatched groups. In other words, people can only be in one group. Don't let the name fool you. ANOVA is the abbreviation for *Analysis of variance* (ANOVA) but it focuses on mean and not variance comparisons. It is called the *analysis of variance* because variances are used to decide whether the means between the two groups are different. How well the variation within group translates into variation between groups is assessed by the p-values are generated using the F-ratio. See p. 279 of ***Elementary Statistics in Social Research***. *Know why the sum of squares cannot be a good measure for variation. Thus, we need to use the mean square because the two sources of variability (between & within) are adjusted by their respective degrees of freedom. This is done by examining how variation within group translates into variation between groups by taking into account the number of individuals in each group. See the top of p. 286 of <u>Elementary Statistics in Social Research</u>.*

The F-ratio is essentially the ratio of the mean of the squares between to the size of between group mean squares to the size of within group mean square:

$$F = \frac{MS_{between}}{MS_{within}}$$ (see p. 287 of ***Elementary Statistics in Social Research***)

As you can see, the F-ratio is essentially a ration between two variances adjusted by their respective degrees of freedom. The p-value depends on how the differences among the different group means compared to the variability of the data within groups (see p. 288 of ***Elementary Statistics in Social Research***). People rarely do ANOVA by hand nowadays, the computer will do it for you.

From p. 252-258 and p. 290-293 of ***Elementary Statistics in Social Research***, you probably noticed that the null hypothesis for ANOVA only tests for no differences among the group means (i.e. equality of group means). Examples for the two-tailed tests on p. 290-293 only focused on two group comparisons (i.e. males and females). However, we know that ANOVA can be used to compare the means between two or more groups. If the F-ratio is large and the p-value is small ($<=0.05$), we know that the means are not equal and null hypothesis is rejected. In other words, the F-ratio only tells us that at least one group means differs from the others but it does not *specifically* tell us which of the groups (i.e. pairs) differ. Once we find out that the F-ratio is large and significant ($P <= 0.05$), we proceed to conduct post-hoc tests like the *Tukey's* HSD to determine which groups differ. A false positive conclusion can be avoided with post-hoc testing because the inflated probability of a Type I error are adjusted. *Pay particular attention to p. 294 of Elementary Statistics in Social Research.* The t-test is essentially a special kind of ANOVA and ANOVA can be regarded as an extension of t-test.

Chapter 8: Chi-Square Test

The *t*-test and the F-test mentioned in the previous weeks are considered parametric tests. Parametric tests assume that samples come from populations that are normally distributed. On the other hand, nonparametric tests relax the restrictions imposed by parametric assumptions and we do not need to have any idea about the functional form of the variables or the data. The Chi-Square χ^2 test is an example of a nonparametric test. Chi-Square test is particularly well-suited when the question of differences can be framed in terms of proportions.

The Chi-Square χ^2 test has two purposes: 1) to test the goodness-of-fit and 2) to test independence.

As a goodness-of-fit test, the Chi-Square (χ^2) can be used to assess how closely the observed frequencies or percentages matches the expected frequencies or percentages under the terms of the null hypothesis. The expected frequency or percentage can be based on previously established distribution obtained from a legitimate source like a government or international agency. See p. 321 of your text. The chi-square goodness-of-fit test is also called the one-way Chi-Square test. See pp. 320-325 of your text. For example, the Census website indicates that 34.8 percent of individuals have income $75,000 or greater (see http://factfinder.census.gov/faces/tableservices/jsf/pages/productview.xhtml?pid=ACS_13_5YR _DP03&src=pt). This is the data we would expect to obtain. However, the data we actually observed from the CBS News/New York Times National Poll indicates that 34.9 percent of individuals have income $75,000 or greater.

income

		Frequency	Percent	Valid Percent	Cumulative Percent
Valid	Under $15,000	74	6.8	7.3	7.3
	$15,000-$30,000	157	14.4	15.6	22.9
	$30,000-$50,000	195	17.9	19.4	42.3
	$50,000-$75,000	230	21.1	22.8	65.1
	Over $75,000	351	32.2	34.9	100.0
	Total	1007	92.5	100.0	
Missing	Won't specify/Refused	82	7.5		
Total		1089	100.0		

The null hypothesis would be no significant difference between the percentages obtained from the Census website and the poll. The research hypothesis would be there is a significant difference between the percentages obtained from the Census website and the poll. The chi-square goodness-of-fit test can be used to assess whether the percentage difference between the Census website and the CBS News/New York Times National Poll are due merely to chance, or other factors (something other than chance) may be at work to cause such difference. The formula of the Chi-Square statistic is

$$\chi^2 = \sum \frac{(f_o - f_e)}{f_e}$$ where f_o is the observed frequency and f_e is the expected frequency (see p. 322 of you text).

The Chi-Square (χ^2) test of independence is used when we want to we have two categorical variables and we want to determine whether these two variables are independent from one another. In other words, it is used to determine whether there is a significant association between the two variables. We could use a chi-square test for independence to determine whether social class is related to voters' registration. The Chi-Square test of independence is also called the two-way Chi-Square test (see p. 325-343).

Regardless of whether the Chi-Square is used as a test of the goodness-of-fit or independence, the expected frequency count for each cell of in the contingency table should be at least 5. The low expected frequency in the denominator will cause the results to be biased or unstable. See p. 338-339 of your text. There are two ways to rectify this problem: 1) combine the categories to increase the expected frequencies and 2) use the Yate's correction. See p. 339 of you text.

Chi-Square Two-Way Test of Significance

Here are all the observed frequencies, f_o's:

Highest grade or level of school completed * Level of damage Crosstabulation

Count

		Level of damage			
		Least damage	Medium damage	Most damaged	Total
Highest grade or level of school completed	Less than 8th grade	13	24	30	67
	9th to 12th grade--no diploma	26	15	23	64
	High school graduate or GED	25	21	31	77
	Some college or 2-year degree	26	15	29	70
	Bachelor's degree	26	7	21	54
	More than bachelor's degree	27	10	14	51
Total		143	92	148	383

<u>Step 1</u>, let's calculate the expected frequency for each cell:

Highest grade or level of school completed	Least damage	Medium damage	Most damaged
Less than 8th grade	$f_e = \dfrac{(67)(143)}{383} = 25.02$	$f_e = \dfrac{(67)(92)}{383} = 16.09$	$f_e = \dfrac{(67)(148)}{383} = 25.89$
9th to 12th grade--no diploma	$f_e = \dfrac{(64)(143)}{383} = 23.90$	$f_e = \dfrac{(64)(92)}{383} = 15.37$	$f_e = \dfrac{(64)(148)}{383} = 24.73$
High school graduate or GED	$f_e = \dfrac{(77)(143)}{383} = 28.75$	$f_e = \dfrac{(77)(92)}{383} = 18.50$	$f_e = \dfrac{(77)(148)}{383} = 29.75$
Some college or 2-year degree	$f_e = \dfrac{(70)(143)}{383} = 26.14$	$f_e = \dfrac{(70)(92)}{383} = 16.81$	$f_e = \dfrac{(70)(148)}{383} = 27.05$
Bachelor's degree	$f_e = \dfrac{(54)(143)}{383} = 20.17$	$f_e = \dfrac{(54)(92)}{383} = 12.97$	$f_e = \dfrac{(54)(148)}{383} = 20.87$
More than bachelor's degree	$f_e = \dfrac{(51)(143)}{383} = 19.04$	$f_e = \dfrac{(51)(92)}{383} = 12.25$	$f_e = \dfrac{(51)(148)}{383} = 19.71$

<u>Step 2</u>, let's calculate the difference between observed and expected frequency for each cell:

Highest grade or level of school completed	Least damage	Medium damage	Most damaged
Less than 8th grade	$f_o - f_e = 13 - 25.02 = -12.02$	$f_o - f_e = 24 - 16.09 = 7.91$	$f_o - f_e = 30 - 25.89 = 4.11$
9th to 12th grade--no diploma	$f_o - f_e = 26 - 23.90 = 2.1$	$f_o - f_e = 15 - 15.37 = -0.37$	$f_o - f_e = 23 - 24.73 = -1.73$
High school graduate or GED	$f_o - f_e = 25 - 28.75 = -3.75$	$f_o - f_e = 21 - 18.50 = 2.5$	$f_o - f_e = 31 - 29.75 = 1.25$
Some college or 2-year degree	$f_o - f_e = 26 - 26.14 = -0.14$	$f_o - f_e = 15 - 16.81 = -1.81$	$f_o - f_e = 29 - 27.05 = 2.05$
Bachelor's degree	$f_o - f_e = 26 - 20.17 = 5.83$	$f_o - f_e = 7 - 12.97 = -5.97$	$f_o - f_e = 21 - 20.87 = 1.87$
More than bachelor's degree	$f_o - f_e = 27 - 19.04 = 7.96$	$f_o - f_e = 10 - 12.25 = -2.25$	$f_o - f_e = 14 - 19.71 = -5.71$

<u>Step 3</u>, let's square the difference between observed and expected frequency for each cell:

Highest grade or level of school completed	Least damage	Medium damage	Most damaged
Less than 8th grade	$(f_o - f_e)^2 = (-12.02)^2 = 144.48$	$(f_o - f_e)^2 = (7.91)^2 = 62.57$	$(f_o - f_e)^2 = (4.11)^2 = 16.89$
9th to 12th grade--no diploma	$(f_o - f_e)^2 = (2.1)^2 = 4.41$	$(f_o - f_e)^2 = (-0.37)^2 = 0.14$	$(f_o - f_e)^2 = (-1.73)^2 = 2.99$
High school graduate or GED	$(f_o - f_e)^2 = (-3.75)^2 = 14.06$	$(f_o - f_e)^2 = (2.5)^2 = 6.25$	$(f_o - f_e)^2 = (1.25)^2 = 1.56$
Some college or 2-year degree	$(f_o - f_e)^2 = (-0.14)^2 = 0.02$	$(f_o - f_e)^2 = (-1.81)^2 = 3.62$	$(f_o - f_e)^2 = (2.05)^2 = 4.20$
Bachelor's degree	$(f_o - f_e)^2 = (5.83)^2 = 33.99$	$(f_o - f_e)^2 = (-5.97)^2 = 35.64$	$(f_o - f_e)^2 = (1.87)^2 = 3.50$
More than bachelor's degree	$(f_o - f_e)^2 = (7.96)^2 = 63.36$	$(f_o - f_e)^2 = (-2.25)^2 = 5.06$	$(f_o - f_e)^2 = (-5.71)^2 = 32.60$

Step 4, let's divide the squared differences by the expected frequencies:

Highest grade or level of school completed	Least damage	Medium damage	Most damaged
Less than 8th grade	$\dfrac{(f_o - f_e)^2}{f_e} = \dfrac{144.48}{25.02} =$ 5.77	$\dfrac{(f_o - f_e)^2}{f_e} = \dfrac{62.57}{16.09} =$ 3.89	$\dfrac{(f_o - f_e)^2}{f_e} = \dfrac{16.89}{25.89} =$ 0.65
9th to 12th grade--no diploma	$\dfrac{(f_o - f_e)^2}{f_e} = \dfrac{4.41}{23.90} =$ 0.18	$\dfrac{(f_o - f_e)^2}{f_e} = \dfrac{0.14}{15.37} =$ 0.01	$\dfrac{(f_o - f_e)^2}{f_e} = \dfrac{2.99}{24.73} =$ 0.12
High school graduate or GED	$\dfrac{(f_o - f_e)^2}{f_e} = \dfrac{14.06}{28.75} =$ 0.49	$\dfrac{(f_o - f_e)^2}{f_e} = \dfrac{6.25}{18.50} =$ 0.34	$\dfrac{(f_o - f_e)^2}{f_e} = \dfrac{1.56}{29.75} =$ 0.05
Some college or 2-year degree	$\dfrac{(f_o - f_e)^2}{f_e} = \dfrac{0.02}{26.14} =$ 0.00	$\dfrac{(f_o - f_e)^2}{f_e} = \dfrac{3.62}{16.81} =$ 0.22	$\dfrac{(f_o - f_e)^2}{f_e} = \dfrac{4.20}{27.05} =$ 0.16
Bachelor's degree	$\dfrac{(f_o - f_e)^2}{f_e} = \dfrac{33.99}{20.17} =$ 1.69	$\dfrac{(f_o - f_e)^2}{f_e} = \dfrac{35.64}{12.97} =$ 2.75	$\dfrac{(f_o - f_e)^2}{f_e} = \dfrac{3.50}{20.87} =$ 0.17
More than bachelor's degree	$\dfrac{(f_o - f_e)^2}{f_e} = \dfrac{63.36}{19.04} =$ 3.33	$\dfrac{(f_o - f_e)^2}{f_e} = \dfrac{5.06}{12.25} =$ 0.41	$\dfrac{(f_o - f_e)^2}{f_e} = \dfrac{32.60}{19.71} =$ 1.65

Step 5, let's sum up all the squared differences:

$$\chi^2 = \Sigma \frac{(f_o - f_e)^2}{f_e} \text{ (see p. 335 of your text)}$$

= 5.77 + 0.18 + 0.49 + 0.00 + 1.69 + 3.33 + 3.89 + 0.01 + 0.34 + 0.22 + 2.75 + 0.41 + 0.65 + 0.12 + 0.16 + 0.17 + 1.65

= 21.83

df = (r − 1)(c − 1) = (6 − 1)(3 − 1) = (5)(2) = 10

Obtained $\chi^2 = 21.83$

Table $\chi^2 = 18.307$

$\alpha = 0.05$

Because obtained χ^2 > Table χ^2, we can reject the null hypothesis and conclude that we found statistically significant evidence to indicate that the level of damage differs due to the level of education. In other words, people with different levels of education experience different levels of damage. Thus, we can conclude that there's an association between the level of damage and the level of education.

How to perform Two-Way Chi-Square Test of Significance in SPSS:

1. Go to Analyze Menu
2. Click on Descriptive Statistics
3. Select Crosstabs
4. A dialog box will open up
5. Select the variables for row and column
6. Click on Statistics
7. Put a check mark beside Chi-Square
8. Click Continue
9. Click OK

Chapter 9: Correlation

As social scientists, our purpose is to isolate, refine and explain the relationship between key variables when we are trying to understand the nature of reality. Correlation is a measure of association. Correlation means that there is a mutual association between two variables. In other words, the values of the two variables change together, either in the same or opposite direction. Both variables are measured at the same time. The purpose of correlation is to measure and describe this relationship. There are two aspects to correlation – direction and strength. In other words, correlation assesses the strength and direction of two variables. It seeks to find out how changes (increase or decrease) in one variable lead to the change in another variable.

The strength and direction of the relationship between two variables are manifested in the correlation coefficients. The value of a correlation coefficient can range from anywhere between -1 to 1. Two variables are positively associated if they move in the same direction. Two variables are negatively associated if they move in the opposite direction. When we talk about correlation coefficients, always think in terms of absolutes. Values close to -1 (-0.6 and above) indicate strong negative relationship between two variables. Values close to 1 (0.6 and above) indicate strong positive relationship between the two variables. See p. 370 of your text. As you can probably see, the sign of the correlation coefficient indicates the direction of the relationship between two variables. The Pearson correlation coefficient assesses the strength and direction of two continuous variables (see p. 371-382 of *Elementary Statistics in Social Research*).

Correlation and Causation

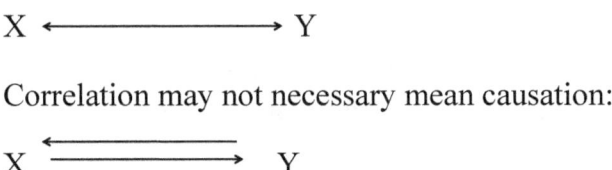

Correlation may not necessary mean causation:

Causation cannot be inferred from correlation. So, how do we determine if there's a causal relationship between the two variables?

There are three ways to do it:

1. We have to show that there is indeed a relationship or covariance between variables. In other words, you want to be able to show that we want to show that a change in the value of one variable will cause the value of another variable to change as well. If one variable is related to another variable, its value will increase or decrease in some predictive manner, along with the increase or decrease in the value of another variable. For example, highly educated individuals are generally healthier. Indeed, a distinguished

legacy of research has also demonstrated that education exerts a long-term and cumulative effect on health by conferring vital resources such as income, wealth, occupational status, knowledge of health risks, and other social-psychological resources (e.g. Mirowsky & Ross, 1998; Ross & Wu, 1995).

Change in one variable  **Change in another**

2. **Time Sequence**: When we talk about correlation, we meant variables that happen or change together at the same time. We need to be able to clearly specify the time sequence of the variables. In other words, we have to know which variable came before the other in order to determine which is the independent variable and which is the dependent variable. A cause must precede the effect. In general, logic, common sense, or reasoning will tell us which is the cause and effect. For instance, logic and common sense will tell us that criminal history of the parents will predict the criminal history of their offspring, not the other way around.

3. **Holding constant all competing causal factors**: Empirically, we can only truly determine whether there is a causal relationship between the two variables if we isolate the effects of the independent variable from the other competing causal factors (control variables). We eliminate the effects of other competing causal factors by holding constant all these competing factors that may also potentially affect the dependent variable. Holding constant means we proceed by finding out whether a particular relationship between two variables looks the same under various conditions by using one or more control variables. Failure to isolate or eliminate the effects of other competing factors could invalidate the results.

Just focus on point # 2 for now. We will talk about points 1 and 3 when we talk about regression analysis.

Partial Correlation

Partial correlation is a method used to assess the net correlation between two variables after the effects of another variable, or several other variables have been "partialled" out. Partialled out means taken away. In other words, when conducting partial correlation, the shared variance of these two variables with the other variables is removed. See p. 382-388 of *Elementary Statistics in Social Research*. Figure 10.7 on p. 383 provides an excellent illustration of partial correlation. Figure 10.8 is worth checking out too. Partial correlation applies to both parametric and nonparametric measures of correlation.

Parametric Measures of Correlation

Note that the correlation coefficient can be calculated either from deviations (see the bottom of p. 373 and the top of p. 374) or from raw scores (see p. 375 of *Elementary Statistics in Social Research*). The $(X - \overline{X})(Y - \overline{Y})$ column on either p. 373 or p. 374 is the covariance, the total is designated as SP. The XY column on p. 375 is also the covariance. As you can see on p. 374, the correlation coefficient, r, is essentially the covariance of X and Y divided by the square root of the product of the variance of X and the variance of Y. Just know that the t-ratio is used to test the significance of correlation coefficient (see p. 376). You don't need to memorize the formulas on p. 385-388.

Non-Parametric Measures of Correlation

Spearman's rank-order correlation

The Spearman's rank-order correlation is the nonparametric alternative of the Pearson product-moment correlation. Spearman's correlation coefficient, r_s, measures the strength of association between two ordinal variables. See the bottom of p. 444 for a definition of ordinal variables. In addition to the two conditions listed on p. 452, Spearman's rank-order correlation is also used for interval variables when the assumptions of the Pearson correlation cannot be met because we are not sure about whether the data is normally distributed (especially when we suspect that the data is skewed). It can also be used the relationship between the two ordinal variables is monotonic (see figures below). Spearman's correlation is more flexible than Pearson's correlation because it makes no assumption that about the distribution of the data. Thus, it can also be used when one variable is measured on an interval scale and the other is measured on an ordinal scale.

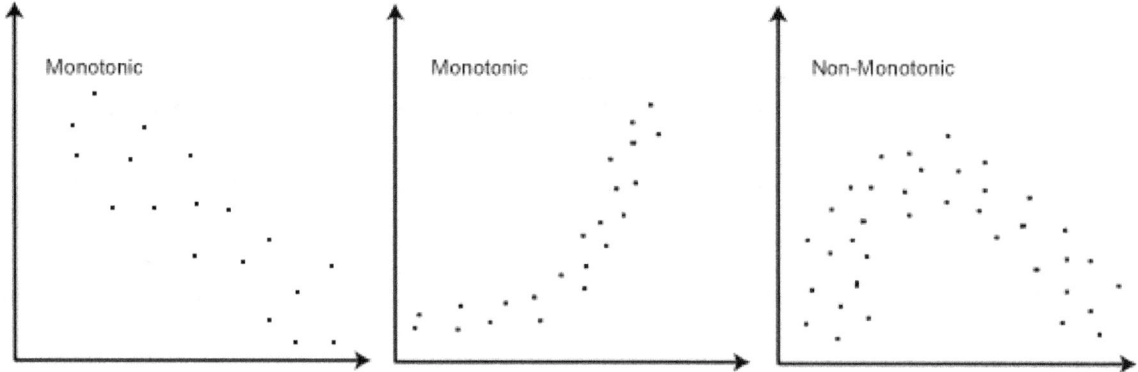

Source: https://statistics.laerd.com/statistical-guides/spearmans-rank-order-correlation-statistical-guide.php

Also, p. 11-15 of *Parametric and nonparametric measures of correlations.pdf* clearly explains when Spearman's correlation coefficient is identical to Pearson's correlation coefficient and when Spearman's correlation coefficient does not match Pearson's correlation coefficient.

Goodman's and Kruskal's Gamma

Pay particular attention to the bottom of p. 452 of **Elementary Statistics in Social Research** to know when it's appropriate to use Goodman's and Kruskal's gamma.

You can also manually calculate the *gamma* by following the steps listed on p. 454-455. They steps should be pretty self-explanatory.

Just know that z-score is used to test the significance of *gamma*, you don't need to memorize the formula on p. 456.

Chi-Square as a Measure of Association

We talked about how the Chi-Square χ^2 test has two purposes: 1) to test the goodness-of-fit and 2) to test independence. We also talked about how the Chi-Square (χ^2) test of independence can be used to determine whether there is a significant association between the two variables. However, the Chi-Square test of independence does NOT tell us the degree of such association.

The phi coefficient is a measure of the degree of association between two nominal variables. Nominal variables are variables with only two categories (e.g. males/females, yes/no, smokers/nonsmokers, etc.). The significance of the phi coefficient is determined by comparing the obtained and Table Chi-Square values[2] (see bottom of p. 458 and Table F on p. 557). Here, the degrees of freedom is (row – 1)(column – 1) or (r -1)(c - 1). There are two rows and two columns on Table 12.5 and therefore the degrees of freedom equals to 1 on p. 458. Using the

---

[2] These are the critical values, the basis on which we make our comparisons.

Chi-Square values; χ^2-test = 3.71 and $\chi^2_{(\alpha/2,1,1)}$ = 3.84 from Chi-Square distribution table on p. 557. Since 3.71 < 3.84 we fail to reject the null hypothesis and conclude that there is no association between smoking status and nationality (see p. 457 of *Elementary Statistics in Social Research*).

Contingency C

Contingency C is an extension of the Chi-Square test when we need to extend beyond a 2 by 2 table. It is used when at least one of our variables has more than two categories. Like the phi coefficient, the significance of Contingency C is determined by comparing the obtained and Table Chi-Square values (see bottom of p. 459 and Table F on p. 557). Here, the degrees of freedom is (row – 1)(column – 1) or (r -1)(c - 1). There are three rows and three columns on Table 12.6 and therefore the degrees of freedom equals to 4 (3 - 1)(3 - 1) on p. 460 of *Elementary Statistics in Social Research*. Using the Chi-Square values; χ^2-test = 7.58 and $\chi^2_{(\alpha/2,2,2)}$ = 9.49 from Chi-Square distribution table on p. 557. Since 7.59 < 9.49 we fail to reject the null hypothesis and conclude that there is no association between child discipline practices and the type of relationship (see p. 459).

Cramer's V

Pay attention to the bottom of p. 460 of *Elementary Statistics in Social Research* that states the limitations of Contingency C. Also, pay attention to p. 461 that states why Cramer's V is a more appropriate measure.

Note that Contingency C can only be used when the number of rows equal the number of columns (2 x 2, 3 x 3, 4 x 4, 5 x 5, and so on). Cramer's V can be used when the number of rows equal the number of columns are different (see the top of p. 461). For instance, we have

Point-Biserial Correlation

Often times, in the social sciences, the choice of variables includes a combination of nominal and interval-ratio scaled variables. When that's the case, a Point-Biserial correlation, a special case of the Pearson product moment correlation, is conducted. You can calculate the Point-Biserial correlation manually using this formula:

$$r_{biserial} = (X_1 - X_0)\left(\frac{pq}{X}\right)\sigma_X$$

Let say that we are interested in finding out whether gender is associated with 8th grade math achievement. Here, we have two variables, *math scores* and *sex*. In this case, X would be *math scores* and Z would be *sex*. Sex is classified as 1=males and 0=females. Thus, X_1 would be the average math scores for males and X_0 would be the average math scores for females. X would be the average math scores for male and female students combined. q would be the proportion of male students in 8th grade and q would be the proportion of female students in 8th grade. Note that $q = 1 - p$.

The average score (males and female students combined) is 85. The average scores for male students is 90. The average scores for female students is 80. The standard deviation (males and female students combined) for the math test is 9. There are 50 students in class (30 females and 20 males). Thus, the proportion of male students (p) would be $\frac{20}{50} = 0.4$ and the proportion male students (q) would be $\frac{30}{50} = 0.6$. Thus, the biserial correlation would be

$$r_{biserial} = (X_1 - X_0)\left(\frac{pq}{X}\right)\sigma_X = (90-80)\left(\frac{0.4*0.6}{85}\right)9 = 0.254$$

The point-biserial correlation between math scores and gender yields a value of 0.254, indicating a weak relationship between the two variables.

Covariance vs. Correlation

Both covariance and correlation measure the degree of association between two variables. Variance is the covariance of the variable with itself, i.e. $(X - \overline{X})(X - \overline{X})$ or (X*X), depending on whether it is calculated either from deviations or raw scores. Both covariance and correlation indicate how the two variables are associated. Positive values of covariance and correlation indicate that the two variables move in the same direction. Negative values of covariance and correlation indicate that the two variables move in the opposite direction. However, unlike correlation, covariance does not tell us the degree of association between these two variables. In other words, covariance does not tell us the degree to which the two variables tend to move together.

Often times, variables have different units of measurement. One variable may be measured in dollars and cents, while the other is measured in something else, be it pound, inch, miles, feet, yards, and so on. Thus, it is impossible to measure the degree of association between the two variables because their units of measurement are not standardized. To measure the degree to the degree of association between two variables, we need to use correlation.

Correlation is actually a standardized covariance. It standardizes the measure of interdependence between two variables and, consequently in its denominator, by taking the square root of the product of the variance of X and the variance of Y. Thus, correlation can tells you whether the two variables are moving in the same or different directions. Because of such standardization, the correlation coefficient, r, will always take on a value between 1 and $-$ 1. Larger covariance will give rise to large correlation coefficients.

Chapter 10: Regression

Regression analysis is the mainstay of social science and is one of the most commonly used statistical techniques. Regression analysis is essentially a statistical process for assessing how two or more variables relate to one another. It often focuses on the potential linear associations between a dependent variable (outcome) and one or more independent variables (or predictors). In other words, the relationship between the two variables can be approximated by a straight line. We can also say that a straight line provides a good linear relationship between the two variables.

We build regression models for inference purposes or predictive purposes. The inference and predictive purposes are normally done through different forms of regression analysis (e.g. ordinary least squares regression, logistic regression, autoregressive models, etc.).

Regression analysis helps one understand how the typical value of the dependent variable (or 'outcome' variable) changes when we change the independent (or 'predictor') variables, while the other independent variables are held constant (fixed). The meaning of holding constant or controlling for will be detailed in the next chapter.

What to do before fitting a regression model?

Before we start fitting regression models, we begin the model building process by performing an Exploratory Data Analysis (EDA). One way to perform an EDA would be to do a scatter plot of dependent variable versus the independent variable(s) can help determine whether the nature of relationship between the two variables is linear. A scatter plot of dependent variable versus the independent variable(s) is called the scatter plot matrix.

As you can see from the figure below, it seems that there is a negative relationship between X and Y. The relationship looks linear. It is possible to fit a straight line to the data using the least squares method (the smallest possible sum of squared error).

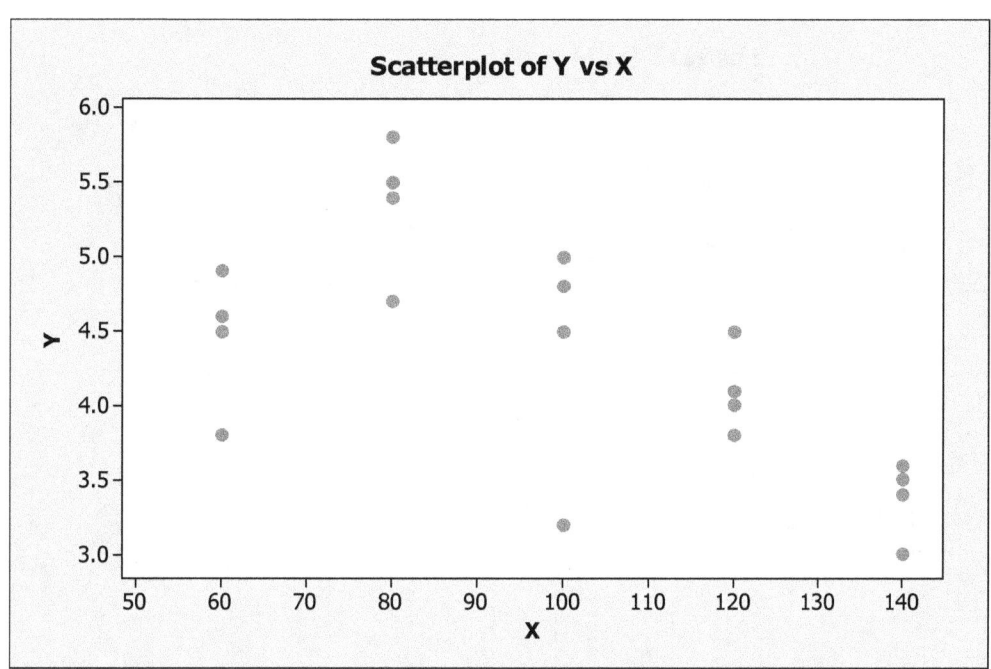

As you can see from the scatter plot matrix below, there is a positive relationship between Y and X3 and Y and X4 and a negative relationship between Y and X8. The scatter plot matrix indicates that the linear model fits the data well (scatter plots) and therefore the linear regression model can be used for prediction.

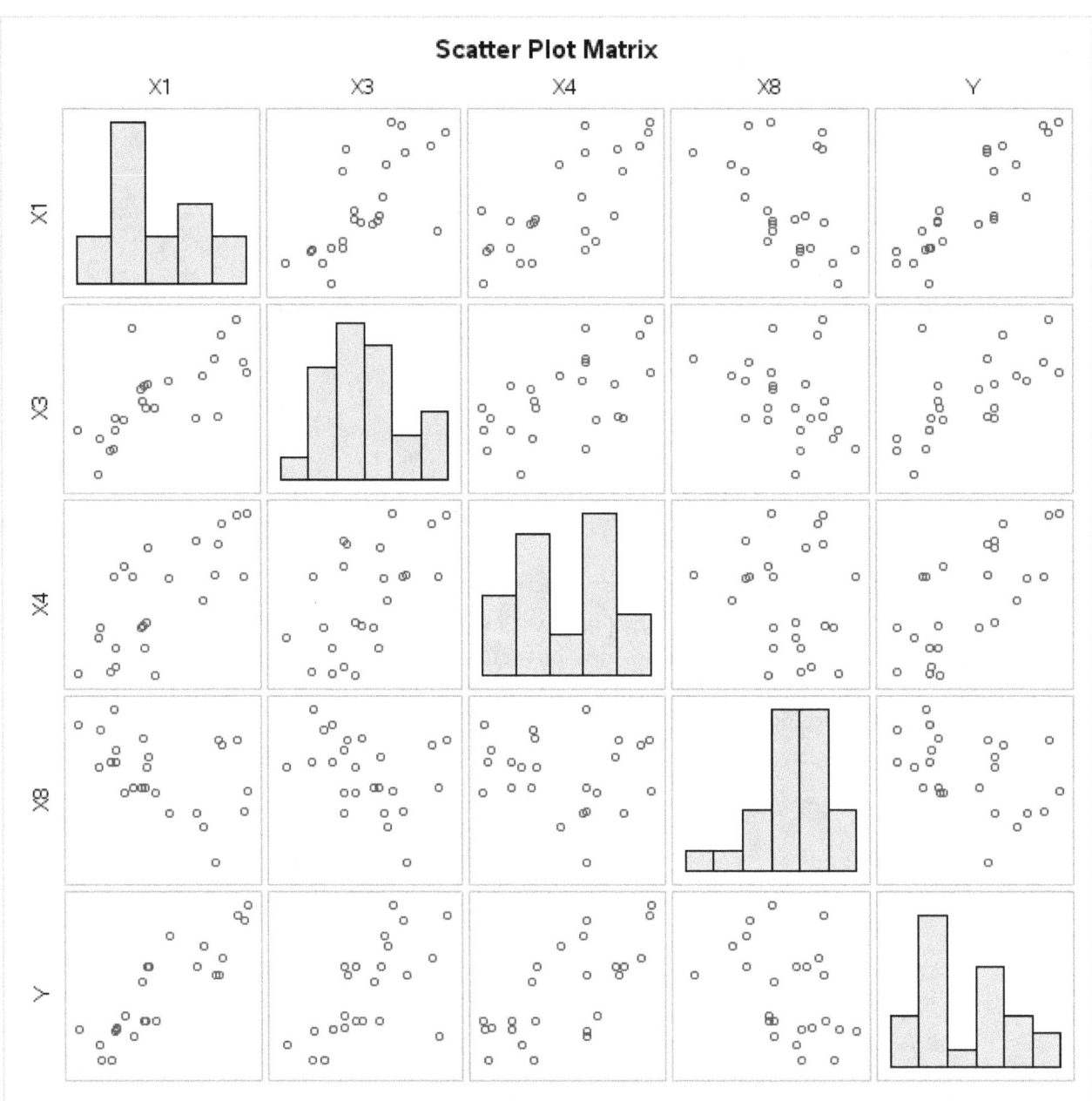

Model formulation

Regression is an approach for modeling the relationship between a continuous dependent variable, denoted as Y, and one or more independent variables, denoted as X. Simple linear regression is used when we only have one independent variable. However this is rarely the case. Our society is an emerging system. Multiple elements (factors) come together to produce an outcome that is greater than the sum of its parts. Thus, we often have to take multiple factors (i.e. independent variables) into account. Multiple linear regression is used when more than one independent variables are used to explain the dependent variable.

The simple linear regression is formulated as:

$Y = a + bX + e$ (see p. 401 of ***Elementary Statistics in Social Research***)

Here, the primary focus of interest is Y, the dependent variable. The value of Y varies from person to person throughout the dataset. However, it is possible for two or more persons with the same Y values. Capital letter X represents the vector[3] of independent variables. In the case of simple linear regression, there's only one item in the vector because there's only one variable. The value of X also varies from person to person but it is also possible for two or more persons with the same X values. X takes on different values and so Y takes on different values. Such variation among different persons is used to explain why the values of Y vary from person to person. In other words, we are implying that X "triggers" Y to take on certain values. If values of X don't vary, it is impossible to do regression analysis. Think about this. When there's no differences in X, X cannot be used to explain the difference (variations) in Y. Movement of X (*via* the different values) generates the regression line or equation. Thus, Y is the explained while X is doing the explaining.

Finally, *e* is the error or disturbance term, which represents the difference between the actual values and the values predicted by the regression model. In other words, we are trying to find the relationship between X and Y knowing that error exists. It is possible for error to exist because we can only estimate the intercept and slope(s) from the sample and we may never know the values for the true intercept and slope(s) in the population. Think back to the M & M's example. See p. 411 of ***Elementary Statistics in Social Research***. In other words, error terms will always exist in any regression analysis because not all variables can be explicitly identified and some variables are omitted in equation or regression model. See p. 408-412 of ***Elementary Statistics in Social Research***. Because of that, any predictions we made using the regression for predictive purposes never leads to results that are 100% accurate.

In regression analysis, our goal is to find out how a one-unit change (increase or decrease) in X lead to the change in the expected (predicted) value of Y. However, often times, we cannot observe the true intercept and slope(s) in the population (i.e. the β_0 and β_k). In other words, we don't know the true values for the β's in the population. We can only estimate these underlying parameters from the sample (i.e. b_0 and b_k). So, how are we going to do that? This is done by first finding a straight line that best describes the relationship between the dependent variable and one or more independent variables. For simple regression, the goal is to find a straight line that best describes the relationship between the dependent variable and the sole independent variable. For multiple regression, the goal is to find a straight line that best describes the relationship between the dependent variable and the independent variables. This is the line that we will use as our estimate of the true linear relationship in the population.

Technically, many different straight lines can be drawn. So, how do we go about determining the line that best describes such relationships? The Least Squares Criterion is used to find the straight line that minimizes the distances between actual and predicted values of Y (the dependent variable). Since all predictive values of Y (i.e. $\hat{Y}$) fall on the regression line, we can also say that we should select the line that minimizes the distance between itself and the actual data points. This will be the line we will use as our estimate of the true linear relationship

---

[3] Vector means a collection of

between the dependent and independent variable(s). In the case of simple regression, we use Least Squares Criterion to find the values for the intercept and slope(s) (i.e. b_0 and b_1) that minimizes the sum of squares errors (SSEs). Thus, we can say that the Least Squares Criterion eventually gives rise to the regression equation, $Y = a + bX + e$. In the case of multiple regression, we use Least Squares Criterion to find the values for the intercept and slope(s) (i.e. b_0, b_1, b_2, b_3, ..., b_k) that minimizes the sum of squares errors (SSEs). If SSE = 0, all the data point we observe fall on the regression line and Y will be equal to $\hat{Y}$ (all the observed values will be equal to all the predicted values).

Figure 14.17 Regression Lines: Perfectly Fit vs. Example 14.13

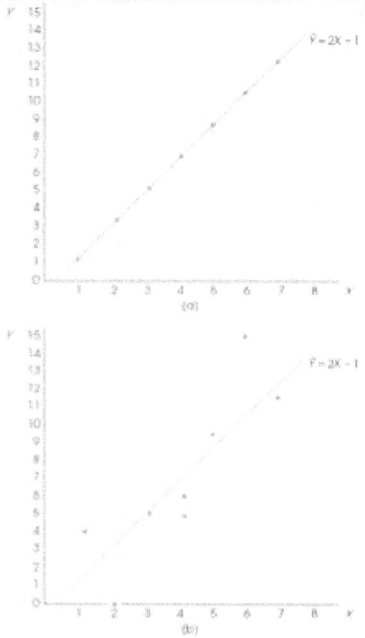

The diagram above looks is a perfect fit. All the data points fall on the line. The diagram below is just another example of a regression line. As the fit gets worse, SSE will get larger.

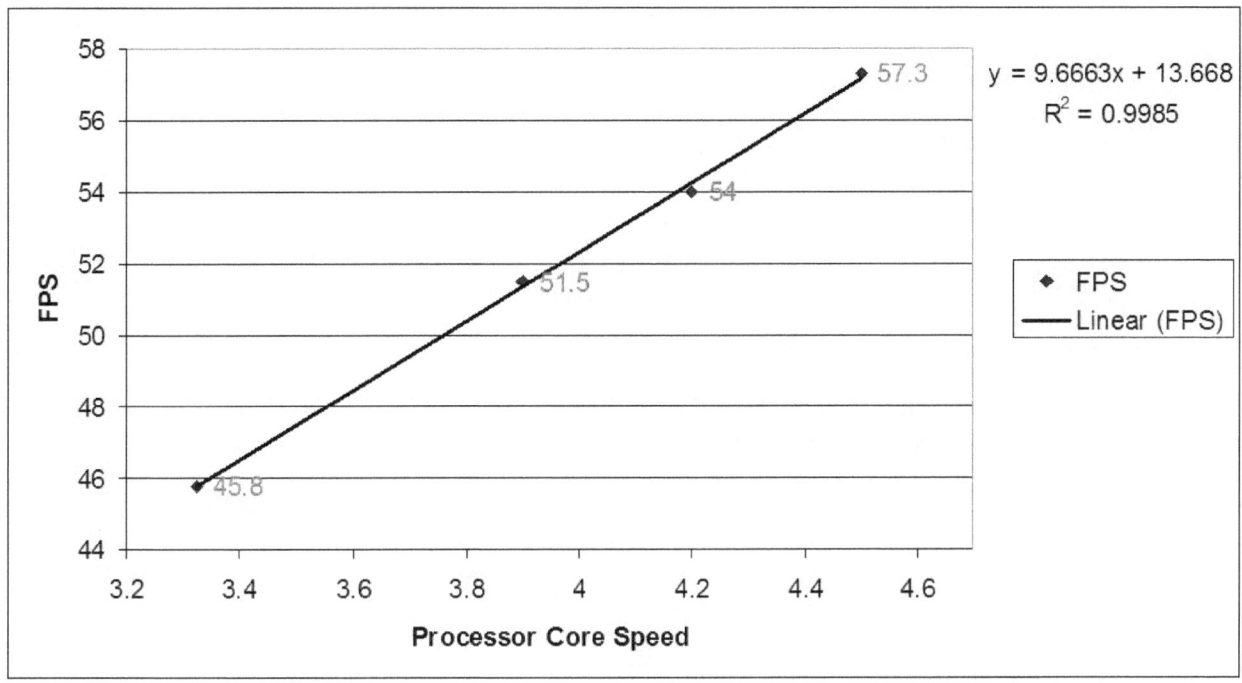

This diagram looks like a close to perfect fit. Almost all data points are on the regression line. The R^2 is very close to 1.

Once we obtained the estimates for the intercept and slopes, we can proceed to substitute a give value of X in order to obtain the predicted value of Y. The result is summarized in the prediction equation as $\hat{Y} = a + bX$ (see p. 405 of **Elementary Statistics in Social Research**). The caret symbol (^) is called "hat" which means prediction. In this case, it's the prediction of Y based on what we know about X (i.e. the values of X). When there are more than two variables, the formula on p. 420 can be extended as $Y = a + b_1X_1 + \ldots + b_kX_k + e$ and the result is summarized in the prediction equation as $\hat{Y} = a + b_1X_1 + \ldots + b_kX_k$. In the case of multiple regression, we are focusing on how the value of Y varies as the values of X_1, X_2, ..., X_k vary.

Regardless of whether it's simple or multiple regression, we hope that $\hat{Y} = a + bX$ and $\hat{Y} = a + b_1X_1 + \ldots + b_kX_k$ will be close to the true line in the population. However, in reality, we never know where the true line lies in the graph. For instance, we know that students' test scores are affected by parent's level of education (measured by years of schooling). Thus, we conducted a regression analysis to see if we increase parent's education by one year increase, student's test scores will increase by _____ points.

In summary, we can also say that

$Y = a + bX + e$ represents the data

$Y = \beta 0 + \beta_k X_k$ represents the real (but unobserved) regression line, and

$\hat{Y} = a + bX$ represents the estimated line.

You don't have to memorize the formulas on p. 420-422.

The definition for dummy variable is explained in the ***Organizing the data chapter*** on the ebook.

Explanations for interaction terms and multicollinearity are pretty clear on p. 425-427 of ***Elementary Statistics in Social Research***.

How should we determine whether an independent variable should be included in the regression model?

The t-statistics can also be used to test the null hypothesis (H_0) that the coefficient is equal to zero (i.e. no relationship between the dependent variable and that particular independent variable). This is done by dividing the estimated coefficient by its standard error. The resulting ratio, which is called the t-statistic, tells us how many standard-error units the coefficient is away from zero. Usually, large t-statistic indicates that the coefficient is statistically different from zero, which means that a linear relationship exists between the dependent variable and that particular independent variable.

The p-value tells us which independent variable(s) is associated with the dependent variable or significantly contributes to the regression model statistically. An independent variable is said to be statistically significant if it is indeed associated with the dependent variable and this association is not merely a chance occurrence (based on the predetermined level of significance). In other words, we have enough evidence to reject the null hypothesis that there is no association between the independent and the dependent variables (B = 0).

The importance of an independent variable can also be assessed by comparing the R^2 for two models, one with that variable excluded, the other with the variable included.

Testing the overall utility/usefulness of the regression model

If we try to assess how useful is the regression model by conducting individual t-tests on each of the β's, we will incorrectly estimate the probability of committing a Type I error and therefore incorrectly reject the null hypothesis (H_0). To test the utility of a regression model, we need a global test like the F-test that tests the β's all at once for overall significance. It tells us whether there's any evidence that any of the independent variables (X_1, X_2, ... X_k) is linearly related to the dependent variable. It is also called the F-ratio (see p. 415 of ***Elementary Statistics in Social Research***).

The null hypothesis (H_0) would be $\beta_1 = \beta_2 = \beta_3 = \ldots \beta_k = 0$.

The research hypothesis (H_a) would be at least one of the β's is not equal to 0 (i.e. $\beta_k \neq 0$), meaning that relationship exist between one of these independent variables and the dependent variable.

We reject H_0 when there's no evidence that at least one of the independent variables is linearly associated with the dependent variable. However, we do not know which predictor is that.

The F-test is the ratio of two variances divided by their appropriate degrees of freedom:

$$\frac{\dfrac{SS_{reg}}{df}}{\dfrac{SS_{error}}{df}} = \frac{\dfrac{R^2}{df}}{\dfrac{1-R^2}{df}} = \frac{MS_{reg}}{MS_{error}} = \frac{Explained\ variance}{Unexplained\ variance}$$ (see p. 415 of ***Elementary Statistics in Social***

Research)

When variances are divided by their appropriate degrees of freedom, we called them mean squared. Thus, we can also say that F-test is the ratio of two mean squared, where the numerator is the explained variance and the denominator is the unexplained variance.

In general, we want the F-ratio to be large that is the mean squared regression (MS_{reg}) to be large with respect to means squared error (MS_{error}). In other words, we want the explained variation to be large with respect to the unexplained variation.

The degrees of freedom for residuals is N – 2 because we lose two degrees of freedom, i.e. $\hat{\beta}_0$ and $\hat{\beta}_1$. It's 8 because you have 10 observations (see p. 402 and 415 ***Elementary Statistics in Social Research***).

The degrees of freedom for regression are determined by the number of independent variables in the model. It's 1 on p. 415 because you only have one independent variable (i.e. see p. 402 of ***Elementary Statistics in Social Research***).

How do we assess the goodness-of-fit of the predictors that can potentially influence the outcome?

Models are epitomes of reality. Models reduce reality to an amount of information that can be handled. Good models can yield insights into complex situation. We use the R-square to quantify how much uncertainty is reduced when the independent variables are used to explain the variability in the dependent variable, that's why we use the R-square. Remember I mentioned in an earlier paragraph that the value of the independent variable varies from person to person and such variation among different persons is used to explain why the values of dependent variable vary from person to person. Thus, we also say that the R-square is used to quantify how much the proportion of variance in the response variable is accounted for by the model fit.

The R-square can be problematic as a goodness-of-fit measure because the value of the R-square never decreases; it only keep increasing when more predictors are added into the model. As such, adjusted R-square is used and is based on the concept of penalized fit because it adjusts for the sample size and the number of predictors. The formula for Adjusted R-square:

$$1 - \frac{(1-R^2)(N-1)}{N-p-1}$$, where N is the total observations and p is the number of predictors in a

dataset. As such, models with the largest number of predictors will not always be selected. For the adjusted R-square, larger values indicate better fit.

How do we assess the accuracy of the independent variables?

Confidence intervals can be used to assess the accuracy of a regression coefficient. This is done by saying how sure (confident) are we (90%? 95%? 99%?) that the true coefficient lies within the boundaries specified by the confidence interval. That is, how sure or confident are we that the average value of the dependent variable increase or decreases between _____ and _____ for every one unit increase in the independent variable.

What can affect the fit of the model?

Often times, the quality of the data can affect the fit of the model. When we say the quality of the data, we meant whether the information (variables) contained in the dataset is relevant to the topic under investigation and if there is a lot of missing information for a particular variable.

Poorly fitted model can mean that predictor variables can hardly predict the outcome variable because having information about the predictor variables only help little in predicting the value of the outcome variable. At times, a poorly fit model implies that you need to add more predictors to the model. Outliers can sometimes affect the fit of the model.

Interpreting multiple regression output

Coefficients^a

Model		Unstandardized Coefficients		Standardized Coefficients	t	Sig.
		B	Std. Error	Beta		
1	(Constant)	32.698	3.173		10.304	.000
	Background Q6a1 Are you presently:Employed full-time?	1.535	2.203	.014	.697	.486
	Background Q6a2 Are you presently:Employed part-time?	-.322	.981	-.007	-.329	.743
	Background Q6a9 Are you presently:Volunteer full time?	.248	1.414	.004	.176	.861
	Background Q6a10 Are you presently:Volunteer part time?	-1.051	.620	-.037	-1.695	.090
	BELIEF (Perceived Control)	-.994	.071	-.316	-13.962	.000
	ISSUES. Finances (Q4,10,16)	.469	.093	.126	5.059	.000
	ISSUES Social Relationship (Q3,9,15)	1.203	.109	.276	11.046	.000
	Black	-1.223	.546	-.056	-2.241	.025
	Hispanic	.382	.671	.015	.569	.570
	Asian	-2.368	1.108	-.047	-2.138	.033
	Other	.367	.988	.008	.372	.710
	Sex	-1.014	.464	-.046	-2.186	.029
	Income2000To3999	-1.314	.648	-.045	-2.029	.043
	Income4000To5999	-2.116	1.106	-.041	-1.913	.056
	Income6000To7999	2.463	1.998	.025	1.233	.218
	IncomeOver8000	.984	1.233	.016	.798	.425
	Married	-.376	.533	-.015	-.705	.481
	High school	-2.439	.643	-.094	-3.792	.000
	Some College, business or trade school	-2.238	.600	-.103	-3.731	.000
	College and beyond	-2.676	.755	-.098	-3.543	.000
	Minimum (age last birthday, 90)	-.067	.028	-.049	-2.363	.018

a. Dependent Variable: Feelings. CESD (Center for Epidemiologic Studies Depression Scale) Sum of 20 items on scale

We will only focus on independent variables that has low p-values (<= 0.05 rounded to two decimal places) because they are likely to be meaningful additions to our model. These are the variables that are highlighted in red. A low p-value means that the null hypothesis (H_0) is rejected and a slope of 0 would be an exceedingly rare event. In other words, a low p-value means that changes in the independent variable's value are related to changes in the dependent variable.

Note that there is no relationship between the status of employment (full time and part time) and depression. Likewise, there is no relationship between volunteering (full time and part time) and depression.

There is a negative relationship between perceived sense of control and depression. Each additional perceived sense of control tends to result in a 0.994 decrease in depression scores.

There is a positive relationship between the number of financial issues and depression. Each additional financial issue tends to result in a 0.469 increases in depression scores. Likewise, there is a positive relationship between the number of issues pertaining to social relationships and depression. Each additional issue pertaining to social relationship tends to result in a 1.203 increase in depression scores.

Blacks and Asians are less depressed than Whites. The dummy variables for race indicate that the expected depression score for Blacks tends to be 1.223 lower than Whites. Likewise, the expected depression score for Asians tends to be 2.368 lower than Whites. Males are less depressed than females. The dummy variables for gender indicate that the expected depression score for males tends to be 1.014 lower than females.

There is a relationship between income and depression. The dummy variables for income indicate that the expected depression score for individuals whose income fall between $2,000 to $3,999 tends to be 1.314 lower than individuals whose income is below $2,000.

There is also relationship between education and depression. The dummy variables for education indicate that the expected depression score for individuals who completed high school tends to be 2.439 lower than individuals did not finish high school. The expected depression score for individuals with some college education tends to be 2.238 lower than individuals did not finish high school. The expected depression score for individuals with college education or higher tends to be 2.676 lower than individuals did not finish high school

There is a negative relationship between age and depression. Each additional age tend to result in a 0.067 decrease in depression scores.

ANOVA[a]

Model		Sum of Squares	df	Mean Square	F	Sig.
1	Regression	65233.269	21	3106.346	47.969	.000[b]
	Residual	98495.754	1521	64.757		
	Total	163729.023	1542			

a. Dependent Variable: Feelings. CESD (Center for Epidemiologic Studies Depression Scale) Sum of 20 items on scale

b. Predictors: (Constant), Minimum (age last birthday, 90), BELIEF (Perceived Control), Income6000To7999, Asian, Other, Background Q6a1 Are you presently:Employed full-time?, Some College, business or trade school, IncomeOver8000, Sex, Income4000To5999, Background Q6a9 Are you presently:Volunteer full time?, Background Q6a2 Are you presently:Employed part-time?, Black, Income2000To3999, Married, Background Q6a10 Are you presently:Volunteer part time?, ISSUES. Finances (Q4,10,16), High school, ISSUES Social Relationship (Q3,9,15), Hispanic, College and beyond

The degrees of freedom for residuals is N − 22 because we lose twenty one degrees of freedom when we have 21 independent variables and one intercept (constant).

The degrees of freedom for regression is determined by the number of independent variables in the model. In this case, it is 21 because we have 21 independent variables.

Model Summary

Model	R	R Square	Adjusted R Square	Std. Error of the Estimate
1	.631[a]	.398	.390	8.04719

a. Predictors: (Constant), Minimum (age last birthday, 90), BELIEF (Perceived Control), Income6000To7999, Asian, Other, Background Q6a1 Are you presently:Employed full-time?, Some College, business or trade school, IncomeOver8000, Sex, Income4000To5999, Background Q6a9 Are you presently:Volunteer full time?, Background Q6a2 Are you presently:Employed part-time?, Black, Income2000To3999, Married, Background Q6a10 Are you presently:Volunteer part time?, ISSUES. Finances (Q4,10,16), High school, ISSUES Social Relationship (Q3,9,15), Hispanic, College and beyond

The R-square is 0.398, meaning that 39.8% of the variance in depression is explained by the regression of depression of these independent variables. In other words, 39.8% of the variance of depression explained by the knowledge of these independent variables. This also suggests that the regression model has adequate ability to make prediction and using information from these independent variables somewhat improved the model fit.

Chapter 11: Logistic Regression

Why we cannot use OLS regression

Unlike the ordinary least squares (OLS) regression, the dependent variable for logistic regression is constrained to whether an individual has (= 1) or does not have (= 0) a certain characteristic or an attribute.

- Registered voter (Yes/No)

- Obese (=1) or not obese (=0)

- Living below the poverty level (Yes/No)

- Have access to clean water (Yes/No)

- Face housing problems (Yes/No)

- Using birth control (Yes/No)

- Admitted (=1) or rejected (=0)

- Baby will be underweight (=1) or normal weight (=0)

- Satisfied with Obamacare (Yes/No)

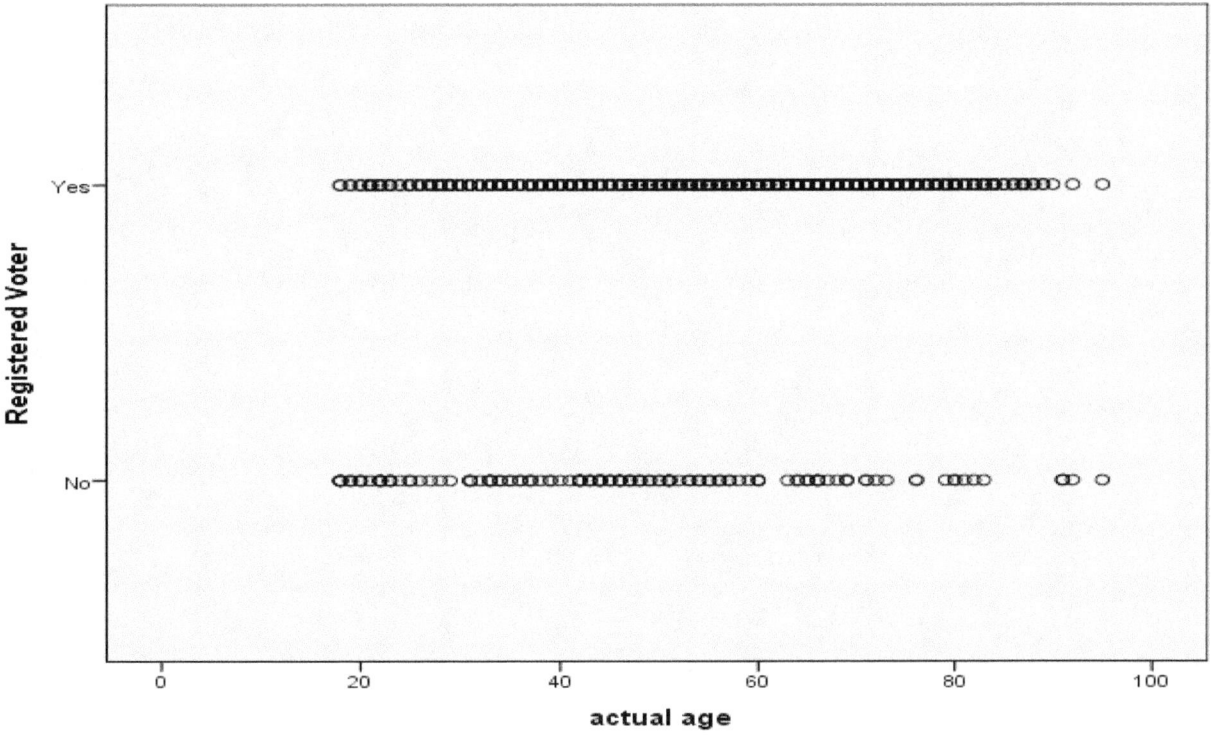

- ❖ We can clearly see the dichotomous nature of the dependent variable in this plot.
- ❖ However, this plot does not provide a clear picture of the nature of the relationship between AGE and voting registration. We only see two straight lines.

❖ Because the dependent variable (Y) take only one of the two values (0 or 1), it is almost meaningless or impossible to predict how a unit change in X affects Y. As such, it can be almost impossible to fit a straight line.

❖ Logistic regression can serve as an alternative to OLS regression. In logistic regression, the dependent variable is called the <u>logit</u>. The <u>logit</u> is NOT a predicted value of Y but it is the probability (ranging from 0-1).

    ↓ If we want to predict the dependent variable based on the information we have about the independent variables, the value for the outcome variable would be a <u>probability</u> ranging from 0 to 1.

❖ If you fit a binary response variable with OLS regression, it would exhibit violations of normality, homogeneity, and out of range values for predicted response.

How is logistic regression formulated?

The probability distribution of the outcome variable follows a Binomial distribution, formulated as:

$P(Y=y) = \pi^y(1-\pi)^{(1-y)}$, $y \in \{0,1\}$

As you can see π is the probability. In this case, y can only take on the value of **0** (the absence of an attribute) or **1** (the presence of an attribute).

The left hand side of the equation is the <u>logit</u>, it is the natural log of the *odds ratio* when the attribute is present, that is, $P(Y = 1)$.

When the attribute is present, $y = 1$ and the probability would be

$$P(Y=1) = \pi^1(1-\pi)^{(1-1)} = \pi$$

When the attribute is absent, $y = 0$ and the probability would be

$$P(Y=0) = \pi^0(1-\pi)^{(1-0)} = 1 - \pi$$

Goodness of Fit

In OLS regression, the R^2 give us a sense the predictive power of the model (i.e. how much uncertainty is reduced when the independent variables are used to explain the variability in the response variable. In logistic regression, we can use the **Nagelkerke's R^2** to assess the overall fit of the model (see p. 433 of ***Elementary Statistics in Social Research***). Other times, the **AIC** is also used.

The likelihood ratio test (aka -2 log likelihood) can be used to compare whether one model has a better fit than the other. Comparison can be done between the null model (no predictors) and the model with predictors. Comparison can also be done between the full[4] and reduced (nested)[5] models. The likelihood ratio rest is minus two (-2) times the difference between the likelihood of the two models. It is based on a ratio test that expresses how likely one more model is more feasible than the other model. However, in order to use the likelihood ratio test, the dependent variables in the models have to be the same in order.

---

[4] Models with all variables included.
[5] Models with a subset of the variables included.

Mathematical Relationships

Both probability and odds look at likelihood or chance of a person having an attribute or a characteristic albeit in different ways. Probability expresses the likelihood or chance as # out of ## while odds expresses the likelihood or chance as # to ##. Notice that both probability and odds differ in their denominators. For probability, the denominator would be the total number of individuals in a study.

$$P(Having\ an\ attribute) = \frac{\#\ of\ individuals\ having\ that\ attribute}{Total\ \#\ of\ individuals\ in\ a\ study}$$

For odds, the denominator would be the # of individuals not having that attribute.

$$Odds = \frac{\#\ of\ individuals\ having\ that\ attribute}{\#\ of\ individuals\ without\ that\ attribute}$$

Let's say that we are trying to find out why students at a particular high school engage in underage drinking. The probability of a high student becoming an underage drinker would be

$$P(Being\ an\ underage\ drinking) = \frac{\#\ of\ underage\ drinkers\ in\ that\ particular\ high\ school}{Total\ \#\ of\ students\ in\ that\ particular\ high\ school}$$

The odds of a student becoming an underage drinker would be

$$Odds(being\ an\ underage\ drinker) = \frac{P(being\ an\ underage\ drinker)}{1 - P(being\ an\ underage\ drinker)}$$

Notice how the denominator constitutes the total # of individuals who are NOT underage drinkers. Also take note that the odds is the ratio of two probabilities $(i.e.\ \frac{P(Being\ an\ underage\ drinker)}{P(Not\ partaking\ in\ underage\ drinking)})$.

If the odds is greater than 1, this suggests that it is more likely for a student to engage in underage drinking or there's a greater chance for a student to engage in underage drinking. On the contrary, odds lesser than 1 suggests that it is more likely for a student to stay sober or a student has a greater chance of staying sober. Odds equal to 1 indicates that there's equal chance of engaging in underage drinking or staying sober. See also p. 432 of ***Elementary Statistics in Social Research***.

Let's forget about the underage drinking example for now. Think of a simple classic example like a coin flip. We know that the chance of getting a head is 50% and the chance of getting a tail is also 50%. Thus, we can say that the probability of getting a head is 0.5. The odds of getting a head is $\frac{P(Head)}{1 - P(Head)} = \frac{0.5}{1 - 0.5} = 1$.

From these two examples, we can say that probability expresses the likelihood or chance using numbers that take any values between 0 and 1. On the other hand, odds express the likelihood or chance using numbers that ranges from 0 to infinity. For the first example, using the odds may make more intuitive sense because we want to compare how much larger one probability is to another. See also p. 430 of **Elementary Statistics in Social Research**. For the coin toss example, it doesn't really matter if the probability or the odds is used to describe the chance or likelihood of getting heads *via* a coin flip. Note that a probability of 0 is the same as odds of 0.

In logistic regression, the dependent variable is called the logit and is expressed as the log of odds:

$$L = log_{[70]}(\frac{p}{1-p})$$

Logits can range from $-\infty$ to ∞. Pay special attention to p. 431 of **Elementary Statistics in Social Research**, that is going to help you when you need to interpret the logistic regression coefficients.

$$log\left(\frac{p}{1-p}\right) = b_0 + b_1X_1 + b_2X_2$$

Notice that when antilog is applied to the left and right sides of the equation, the right side of the equation becomes exponentiated. Thus we have, $log\left(\frac{p}{1-p}\right) = e^{b0 + b1X1 + b2X2}$

For Table 11.9 of **Elementary Statistics in Social Research**, you need to pay particular attention to the *b*, *Sig.*, and *Exp(b)* columns. *b*'s are the coefficients expressed in log-odds. By merely looking at *b*'s, we can tell whether the variable has a positive or negative effect on the dependent variables. In this example, the values for b's are positive (0.269) for high school average (HSavg) and negative for Gender. Thus, we can simply conclude that there's a positive association between college admission and high school average or the likelihood of college admission increases with high school average. However, this interpretation is somewhat crude and less intuitive. Thus, we also need to look at the *Exp(b)* column to get a more intuitive interpretation. High school average (HSavg) is the only significant independent variable in that table. When you exponentiate that coefficient ($e^{0.269}$), you get 1.309. Just type =exp(0.269) in Excel and you will see. This suggests that for every one-point increase in high school average, the odds of college admission increase by a factor of 1.309 or 30.9 percent. Pay special attention to p. 432-433 of **Elementary Statistics in Social Research**.

Chapter 12: Ordinal regression

Logistic regression can be extended to handle polytomous dependent variables, that is, dependent variables that have three or more categories. Ordinal logistic regression is used when the polytomous dependent variable has three or more ordered categories (like the example shown below). When we use ordinal logistic regression, we are focusing on the magnitudes of a variable (e.g. opinions about affordable healthcare).

Alternatively, we can make each category a new dummy variable and run a series of logistic regressions. We can also collapse the number of categories to two and perform a logistic regression However, this approach can drastically changes our research question(s) and can also lead to a loss of information (each category may have something unique to contribute).

Case Processing Summary

		N	Marginal Percentage
Opinion of Affordable Health Care Act	Strongly approve	169	21.3%
	Somewhat approve	179	22.5%
	Somewhat disapprove	111	14.0%
	Strongly disapprove	336	42.3%
Respondent's Sex	Male	375	47.2%
	Female	420	52.8%
income	Under $15,000	47	5.9%
	$15,000-$30,000	112	14.1%
	$30,000-$50,000	147	18.5%
	$50,000-$75,000	186	23.4%
	Over $75,000	303	38.1%
Ethnicity	Non-Hispanic White	698	87.8%
	Non-Hispanic Black	62	7.8%
	Asian	12	1.5%
	Other	23	2.9%
Presence of kids	Yes, under 18	173	21.8%
	Yes, over 18	392	49.3%
	No	230	28.9%
Valid		795	100.0%
Missing		294	
Total		1089	

The Case Processing Summary just shows the frequency and % distribution of each levels of a variable.

Pseudo R-Square

Cox and Snell	.076
Nagelkerke	.082
McFadden	.030

Link function: Logit.

		Estimate	Std. Error	Wald	df	Sig.	95% Confidence Interval Lower Bound	95% Confidence Interval Upper Bound
Threshold	Strongly approve	-1.378	.469	8.632	1	.003	-2.298	-.459
	Somewhat approve	-.245	.466	.277	1	.599	-1.159	.668
	Somewhat disapprove	.361	.466	.599	1	.439	-.553	1.274
Location	AGE	-.011	.005	4.782	1	.029	-.021	-.001
	Male	.216	.133	2.645	1	.104	-.044	.477
	Female	0ª	.	.	0	.	.	.
	Under $15,000	-.168	.290	.336	1	.562	-.738	.401
	$15,000-$30,000	.142	.208	.462	1	.497	-.267	.550
	$30,000-$50,000	.244	.190	1.643	1	.200	-.129	.616
	$50,000-$75,000	.021	.173	.014	1	.905	-.319	.361
	Over $75,000	0ª	.	.	0	.	.	.
	Non-Hispanic White	.287	.389	.545	1	.460	-.475	1.049
	Non-Hispanic Black	-1.303	.453	8.284	1	.004	-2.191	-.416
	Asian	-.669	.650	1.060	1	.303	-1.943	.605
	Other	0ª	.	.	0	.	.	.
	Yes, under 18	.344	.191	3.238	1	.072	-.031	.718
	Yes, over 18	.567	.171	10.961	1	.001	.231	.903
	No	0ª	.	.	0	.	.	.

Link function: Logit.

a. This parameter is set to zero because it is redundant.

From the table, you can see that ordinal logistic regression is an extension of the binary logistic regression. Here, the dependent variable has more than two categories.

Unlike logistic regression, each category of the dependent has its own intercept but the coefficients are the same for all categories. The threshold coefficients are merely the intercepts for each category of the dependent variable. They represent the likelihood an individual progress from one category to the next. We usually do not interpret them.

We can interpret the independent variables in terms of likelihood or odds.

Likelihood interpretation:

The table above suggests that age, race and presence of children are related to the perceptions about affordable healthcare. The likelihood of being in a higher disapproval category of affordable healthcare decreases with age (see the negative coefficient). Blacks are less likely than other racial ethnic groups to be in a higher disapproval category of affordable healthcare (see the negative coefficient). Individuals with adult children (over 18) are more likely to be in a higher disapproval category affordable healthcare than their childless counterparts (see the positive coefficient).

Odds interpretation:

Results from the table above indicate that the odds of being in a higher disapproval category decreases with age ($e^{-0.011} = 0.989$). Blacks have lower odds of being in a higher disapproval category of affordable healthcare than other racial ethnic groups ($e^{-1.303} = 0.271$). Individuals with adult children (over 18) have higher odds of being in a higher disapproval category of affordable healthcare than their childless counterparts ($e^{0.567} = 1.763$).

Chapter 13: Multinomial logistic regression

As mentioned earlier, logistic regression can be extended to handle polytomous dependent variables. Ordinal logistic regression is used when the polytomous dependent variable has more than two ordered categories. Multinomial logistic regression is used when the dependent variable has more than two or more unordered categories. In other words, the polytomous dependent variable is nominal (the categories cannot be ordered like the example shown below).

As is the case with ordinal logistic regression, we can make each category a new dummy variable and run a series of logistic regressions. One problem with this approach is that each analysis is potentially run on a different sample and we may end up with a probability of greater than 1 if all possible outcomes are selected. As is the case with ordinal logistic regression, we can also collapse the number of categories to two and perform a logistic regression However, this approach can drastically changes our research question(s) and can also lead to a loss of information (each category may have something unique to contribute). Hence, using multinomial logistic regression may lead to more robust[6] parameter estimates and may reduce the errors sums of squares.

When we use ordinal logistic regression, we are focusing on the magnitudes of a variable (e.g. political philosophy). Nevertheless, as shown in the example below, we do not necessarily have to take the ordering into account even if the categories of the dependent variable can be ordered. Using ordinal logistic regression will make the model more parsimonious if we can really order/rank the dependent variable and the assumption of proportional are also being met (see example below). When we use multinomial logistic regression, we are viewing the categories within a variable as a set of choices. This is appropriate when the structure of the dependent variable is clearly choice-specific.

---

[6] Robustness refers to the ability of a statistical test to provide insight to a problem even when their assumptions are slightly altered or violated.

Results from ordinal logistic regression:

Case Processing Summary

		N	Marginal Percentage
political philosophy	Liberal	194	22.2%
	Moderate	352	40.3%
	Conservative	327	37.5%
Respondent's Sex	Male	393	45.0%
	Female	480	55.0%
income	Under $15,000	59	6.8%
	$15,000-$30,000	127	14.5%
	$30,000-$50,000	159	18.2%
	$50,000-$75,000	205	23.5%
	Over $75,000	323	37.0%
Ethnicity	Non-Hispanic White	754	86.4%
	Non-Hispanic Black	76	8.7%
	Asian	15	1.7%
	Other	28	3.2%
Presence of kids	Yes, under 18	187	21.4%
	Yes, over 18	438	50.2%
	No	248	28.4%
Valid		873	100.0%
Missing		216	
Total		1089	

As mentioned earlier, the Case Processing Summary just shows the frequency and % distribution of each levels of a variable.

Pseudo R-Square

Cox and Snell	.028
Nagelkerke	.031
McFadden	.013

Link function: Logit.

Parameter Estimates

		Estimate	Std. Error	Wald	df	Sig.	95% Confidence Interval Lower Bound	95% Confidence Interval Upper Bound
Threshold	Liberal	-.909	.438	4.295	1	.038	-1.768	-.049
	Moderate	.897	.438	4.187	1	.041	.038	1.756
Location	AGE	-.004	.005	.579	1	.447	-.013	.006
	Male	.262	.129	4.124	1	.042	.009	.515
	Female	0ª	.	.	0	.	.	.
	Under $15,000	.108	.267	.165	1	.685	-.415	.632
	$15,000-$30,000	.169	.200	.712	1	.399	-.224	.562
	$30,000-$50,000	.239	.185	1.672	1	.196	-.123	.602
	$50,000-$75,000	-.013	.168	.006	1	.938	-.342	.316
	Over $75,000	0ª	.	.	0	.	.	.
	Non-Hispanic White	.214	.360	.352	1	.553	-.492	.920
	Non-Hispanic Black	-.520	.413	1.582	1	.208	-1.330	.290
	Asian	-.751	.602	1.559	1	.212	-1.931	.428
	Other	0ª	.	.	0	.	.	.
	Yes, under 18	.242	.186	1.699	1	.192	-.122	.606
	Yes, over 18	.431	.166	6.749	1	.009	.106	.757
	No	0ª	.	.	0	.	.	.

Link function: Logit.
a. This parameter is set to zero because it is redundant.

As is the case with linear regression, we will only focus on independent variables that has low p-values (<= 0.05 rounded to two decimal places) because they are likely to be meaningful additions to our model. These are the variables that are highlighted in red.

Likelihood interpretation:

The table above suggests that sex and presence of children are related to the political philosophy. Males are more likely than females to become more conservative in their political stance (see the positive coefficient). Individuals with adult children (over 18) are more likely to become more conservative in their political stance than their childless counterparts (see positive coefficient).

Odds interpretation:

Results from the table above indicate that males have higher odds of adopting a more conservative political stance than females ($e^{0.262} = 1.299$). Individuals with adult children (over 18) have higher odds of adopting a more conservative political stance than their childless counterparts ($e^{0.431} = 1.539$).

Test of Parallel Lines[a]

Model	-2 Log Likelihood	Chi-Square	df	Sig.
Null Hypothesis	1581.630			
General	1565.854	15.776	11	.150

The null hypothesis states that the location parameters (slope coefficients) are the same across response categories.

a. Link function: Logit.

Results from multinomial logistic regression:

Case Processing Summary

		N	Marginal Percentage
political philosophy	Liberal	194	22.2%
	Moderate	352	40.3%
	Conservative	327	37.5%
Respondent's Sex	Male	393	45.0%
	Female	480	55.0%
income	Under $15,000	59	6.8%
	$15,000-$30,000	127	14.5%
	$30,000-$50,000	159	18.2%
	$50,000-$75,000	205	23.5%
	Over $75,000	323	37.0%
Ethnicity	Non-Hispanic White	754	86.4%
	Non-Hispanic Black	76	8.7%
	Asian	15	1.7%
	Other	28	3.2%
Presence of kids	Yes, under 18	187	21.4%
	Yes, over 18	438	50.2%
	No	248	28.4%
Valid		873	100.0%
Missing		216	
Total		1089	
Subpopulation		627[a]	

a. The dependent variable has only one value observed in 510 (81.3%) subpopulations.

Pseudo R-Square

Cox and Snell	.043
Nagelkerke	.049
McFadden	.021

Parameter Estimates

political philosophy[a]		B	Std. Error	Wald	df	Sig.	Exp(B)	95% Confidence Interval for Exp(B)	
								Lower Bound	Upper Bound
Liberal	Intercept	.055	.618	.008	1	.929			
	AGE	.003	.007	.148	1	.700	1.003	.989	1.017
	Male	-.356	.188	3.580	1	.058	.700	.484	1.013
	Female	0[b]	.	.	0	.	.	.	.
	Under $15,000	-.013	.370	.001	1	.973	.987	.478	2.041
	$15,000-$30,000	-.159	.290	.299	1	.584	.853	.483	1.506
	$30,000-$50,000	-.366	.285	1.653	1	.199	.693	.397	1.212
	$50,000-$75,000	.124	.240	.266	1	.606	1.132	.708	1.810
	Over $75,000	0[b]	.	.	0	.	.	.	.
	Non-Hispanic White	-.296	.510	.338	1	.561	.743	.274	2.019
	Non-Hispanic Black	.759	.593	1.634	1	.201	2.135	.667	6.831
	Asian	.994	.872	1.297	1	.255	2.701	.488	14.934
	Other	0[b]	.	.	0	.	.	.	.
	Yes, under 18	-.295	.264	1.246	1	.264	.745	.444	1.250
	Yes, over 18	-.562	.246	5.228	1	.022	.570	.352	.923
	No	0[b]	.	.	0	.	.	.	.
Moderate	Intercept	.114	.548	.043	1	.836			
	AGE	.013	.006	4.577	1	.032	1.013	1.001	1.026
	Male	-.220	.158	1.923	1	.165	.803	.588	1.095
	Female	0[b]	.	.	0	.	.	.	.
	Under $15,000	-.501	.339	2.185	1	.139	.606	.312	1.178
	$15,000-$30,000	-.437	.247	3.144	1	.076	.646	.398	1.047
	$30,000-$50,000	-.162	.223	.529	1	.467	.850	.549	1.317
	$50,000-$75,000	-.333	.209	2.540	1	.111	.716	.475	1.080
	Over $75,000	0[b]	.	.	0	.	.	.	.
	Non-Hispanic White	-.244	.451	.292	1	.589	.784	.324	1.898
	Non-Hispanic Black	.600	.534	1.261	1	.261	1.821	.640	5.187
	Asian	.462	.840	.302	1	.583	1.587	.306	8.238
	Other	0[b]	.	.	0	.	.	.	.
	Yes, under 18	-.295	.233	1.601	1	.206	.745	.472	1.176
	Yes, over 18	-.510	.207	6.053	1	.014	.600	.400	.901
	No	0[b]	.	.	0	.	.	.	.

a. The reference category is: Conservative.
b. This parameter is set to zero because it is redundant.

As is the case with ordinal logistic regression, we can interpret the independent variables in terms of likelihood or odds.

From the table, you can see that multinomial logistic regression is an extension of the binary logistic regression. As is the case with ordinal logistic regression, the dependent variable has more than two categories. The purpose of ordinal logistic regression is to assess the effects of changes in IVs on the actual approval / disapproval level underlying the DV. Multinomial logistic regression differs from ordinal logistic regression in their interpretations of coefficients. In multinomial logistic regression, the dependent variable is listed as a set of choices.

Likelihood interpretation:

The table above suggests that the presence of children is related to the political philosophy. Moving from being childless (lowest category) to having adult children (the highest category) is associated with a 0.562 decrease in the relative log odds of being a liberal versus being a conservative. The likelihood of adopting a moderate political stance increases as a person ages (see the positive coefficient). A one year increase in age is associated with a 0.13 increase in the relative log odds of being a moderate versus being a conservative. Individuals with adult children (over 18) are less likely to adopt a moderate political stance than their childless counterparts (see the negative coefficient). Moving from being childless to having adult children is associated with a 0.510 decrease in the relative log odds of being a moderate versus being a conservative. In other words, results suggest that people increasingly adopt a more conservative political stance as they themselves and their children grow older.

Odds interpretation:

Results from the table above indicate that moving from being childless to having adult children decreases the odds of being a liberal versus being a conservative ($e^{-0.562} = 0.57$). The odds of adopting a moderate political stance versus a conservative political stance increases as a person ages ($e^{0.013} = 1.013$). Moving from being childless to having adult children decreases the odds of being a moderate versus being a conservative ($e^{-0.51} = 0.6$). As you can see, the odds interpretation is consistent with the likelihood interpretation.

Chapter 14: Two-step clustering

Introduction

Often times, in the social sciences, the choice of clustering variables includes a combination of nominal, ordinal, and interval-ratio scaled measures, clustering techniques that can only accommodate interval-ratio or continuous variables, hierarchical agglomerative clustering and k-means clustering are inappropriate. Hence, two-step clustering within SPSS was employed to segment individuals based on their age, sex, education level, marital status, type of medical facility used, as well as their patterns and preferences of outpatient care usage. This approach consists of a pre-clustering and hierarchical clustering stage. It combines characteristics of k-means clustering and hierarchical agglomerative clustering approaches in a two-step algorithm. In the first stage, a sequential clustering algorithm akin to the k-means approach to scan the observations one by one. Observations are merged to form initial cluster based on a designated distance criterion. In the second stage, hierarchical agglomerative clustering is performed to sequentially combine the initial clusters into optimal number of desired clusters. Goodness-of-fit measures like the Akaike's Information Criterion (AIC) or the Bayes Information Criterion (BIC) are used to determine the number of optimal clusters to retain.

Steps in conducting k-means clustering in SPSS:

1. Go to the Analyze menu.
2. Select Classify.
3. Select TwoStep Cluster…
4. Select the categorical and continuous variables. You can also click on the Output button and check Pivot tables and Create cluster membership variable (see Figure 4.2).

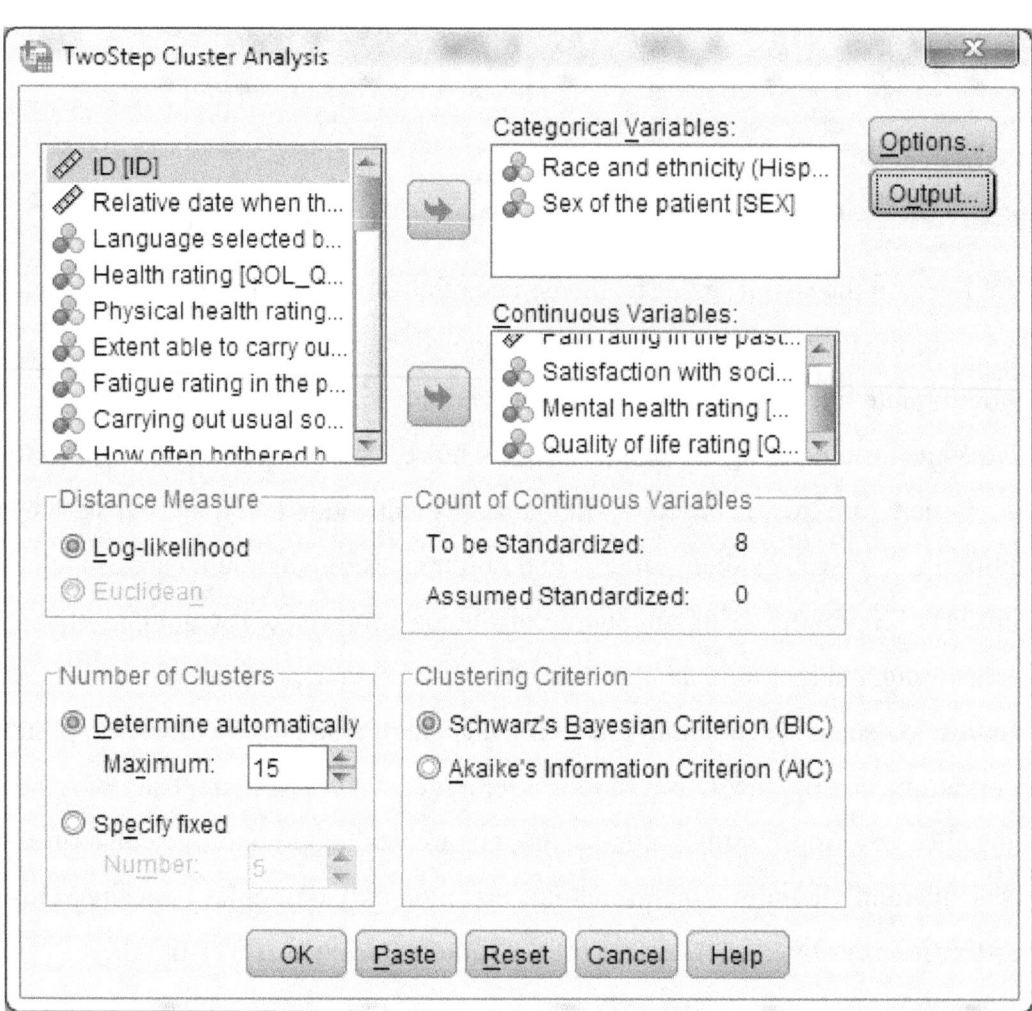

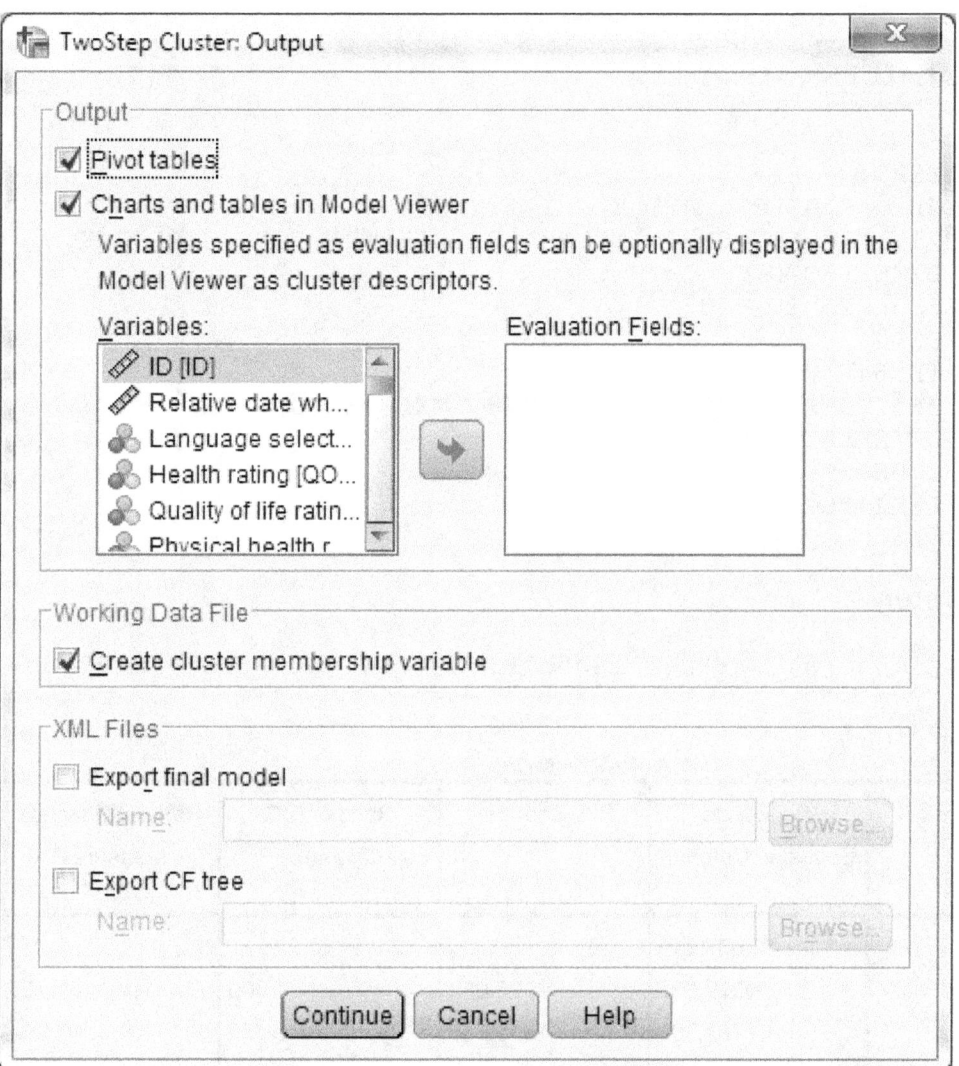

5. Click Continue.
6. Click OK.

Example from 2011-2012 Computer Assisted Quality of Life and Symptom Assessment of Complex Patients (ICPSR 34543)

```
TWOSTEP CLUSTER
  /CATEGORICAL VARIABLES=RACE_ETHNICITY SEX
  /CONTINUOUS VARIABLES=CHRONICDISEASESCORE CHARLSONSCORE AGE
  /DISTANCE LIKELIHOOD
  /NUMCLUSTERS AUTO 15 BIC
  /HANDLENOISE 0
  /MEMALLOCATE 64
  /CRITERIA INITHRESHOLD(0) MXBRANCH(8) MXLEVEL(3)
  /VIEWMODEL DISPLAY=YES
  /PRINT IC COUNT SUMMARY
  /SAVE VARIABLE=TSC_7500.
```

TwoStep Cluster

[DataSet2] C:\Users\User\Documents\Elsevier\34543-0001-Data.sav

Auto-Clustering

Number of Clusters	Schwarz's Bayesian Criterion (BIC)	BIC Change[a]	Ratio of BIC Changes[b]	Ratio of Distance Measures[c]
1	12007.215			
2	9277.533	-2729.682	1.000	1.755
3	7761.765	-1515.768	.555	1.294
4	6610.968	-1150.797	.422	1.700
5	5971.896	-639.072	.234	1.209
6	5459.244	-512.652	.188	1.568
7	5165.556	-293.689	.108	1.392
8	4980.360	-185.196	.068	1.033
9	4803.979	-176.381	.065	1.349
10	4696.934	-107.045	.039	1.074
11	4603.663	-93.271	.034	1.019
12	4513.818	-89.846	.033	1.070
13	4435.885	-77.933	.029	1.242
14	4390.984	-44.900	.016	1.131
15	4361.861	-29.124	.011	1.062

a. The changes are from the previous number of clusters in the table.

b. The ratios of changes are relative to the change for the two cluster solution.

c. The ratios of distance measures are based on the current number of clusters against the previous number of clusters.

Cluster Distribution

		N	% of Combined	% of Total
Cluster	1	527	25.3%	22.4%
	2	720	34.6%	30.6%
	3	381	18.3%	16.2%
	4	454	21.8%	19.3%
	Combined	2082	100.0%	88.4%
Excluded Cases		274		11.6%
Total		2356		100.0%

Cluster Profiles

Variables	Cluster 1 (N = 527)	Cluster 2 (N = 720)	Cluster 3 (N = 381)	Cluster 4 (N = 454)
Chronic Disease Score	1.59 (0.79)	1.47 (0.78)	1.46 (0.79)	1.44 (0.80)
Carlson Score	0.94 (0.88)	0.84 (0.94)	1.34 (2.12)	1.11 (1.01)
Age of the patient at the time of the interview	55.45 (12.41)	62.70 (11.57)	57.40 (10.92)	56.94 (13.02)
Race				
American Indian-Alaskan Native	0.00%	0.00%	7.87%	0.00%
Asian	0.00%	0.00%	34.91%	0.00%
Black	0.00%	100.00%	2.89%	100.00%
Hispanic	100.00%	0.00%	2.89%	0.00%
Native Hawaiians-Pacific Islanders	0.00%	0.00%	1.57%	0.00%
White	0.00%	0.00%	49.87%	0.00%
Sex				
Female	54.84%	100.00%	53.28%	0.00%
Male	45.16%	0.00%	46.72%	100.00%

Cluster 1

The mean age in this cluster is about 55. Individuals in this cluster have relatively high Chronic Disease Score. All individuals in this cluster are Hispanics.

Cluster 2

The mean age in this cluster is about 57. Individuals in this cluster have relatively low Carlson Score. All individuals in this cluster are Black females.

Cluster 3

The mean age in this cluster is nearly 63. Individuals in this cluster have relatively high Carlson Score.

Cluster 4

The mean age in this cluster is nearly 57. All individuals in this cluster are Black males.

Model Summary

Algorithm	TwoStep
Inputs	5
Clusters	4

Cluster Quality

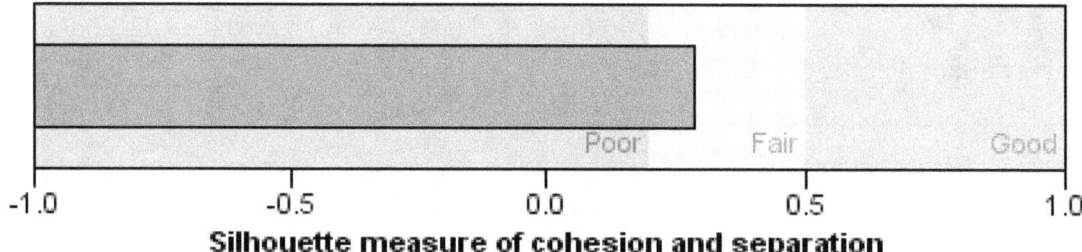

Silhouette measure of cohesion and separation

Example from CBS News/New York Times National Poll, ICPSR 34617

```
TWOSTEP CLUSTER
  /CATEGORICAL VARIABLES=Q4 RACE CENR
  /CONTINUOUS VARIABLES=AGE
  /DISTANCE LIKELIHOOD
  /NUMCLUSTERS AUTO 15 BIC
  /HANDLENOISE 0
  /MEMALLOCATE 64
```

```
/CRITERIA INITHRESHOLD(0) MXBRANCH(8) MXLEVEL(3)
/VIEWMODEL DISPLAY=YES
/PRINT IC COUNT SUMMARY.
```

TwoStep Cluster

[DataSet5] C:\Users\User\Documents\Elsevier\34617-0001-Data.sav

Auto-Clustering

Number of Clusters	Schwarz's Bayesian Criterion (BIC)	BIC Change[a]	Ratio of BIC Changes[b]	Ratio of Distance Measures[c]
1	4527.774			
2	3632.604	-895.169	1.000	1.349
3	2986.491	-646.113	.722	1.315
4	2511.276	-475.215	.531	1.318
5	2167.012	-344.264	.385	1.022
6	1831.741	-335.271	.375	2.352
7	1728.150	-103.591	.116	1.771
8	1699.141	-29.008	.032	1.064
9	1675.945	-23.197	.026	1.107
10	1661.512	-14.433	.016	1.152
11	1657.950	-3.561	.004	1.116
12	1661.817	3.867	-.004	1.108
13	1671.928	10.110	-.011	1.046
14	1684.590	12.662	-.014	1.042
15	1699.476	14.886	-.017	1.004

a. The changes are from the previous number of clusters in the table.

b. The ratios of changes are relative to the change for the two cluster solution.

c. The ratios of distance measures are based on the current number of clusters against the previous number of clusters.

		N	% of Combined	% of Total
Cluster	1	232	26.5%	21.3%
	2	69	7.9%	6.3%
	3	135	15.4%	12.4%
	4	109	12.4%	10.0%
	5	164	18.7%	15.1%
	6	167	19.1%	15.3%
	Combined	876	100.0%	80.4%
Excluded Cases		213		19.6%
Total		1089		100.0%

Cluster Profiles

Variable	Cluster 1 (N = 232)	Cluster 2 (N = 69)	Cluster 3 (N = 135)	Cluster 4 (N = 109)	Cluster 5 (N = 164)	Cluster 6 (N = 167)
Age	58.36 (15.28)	57.35 (17.90)	57.11 (15.91)	53.20 (16.66)	58.02 (14.70)	58.90 (16.77)
Likely to vote in November 2012						
Definitely vote	100.00%	0.00%	100.00%	98.17%	100.00%	100.00%
Probably vote	0.00%	78.26%	0.00%	0.00%	0.00%	0.00%
Probably not vote	0.00%	21.74%	0.00%	1.83%	0.00%	0.00%
Race						
White	100.00%	86.96%	100.00%	0.00%	100.00%	100.00%
Black	0.00%	5.80%	0.00%	64.22%	0.00%	0.00%
Asian	0.00%	0.00%	0.00%	7.34%	0.00%	0.00%
Other	0.00%	7.25%	0.00%	28.44%	0.00%	0.00%
Census Region						
Northeast	0.00%	18.84%	0.00%	23.85%	0.00%	100.00%
North Central	0.00%	30.43%	0.00%	23.85%	100.00%	0.00%
South	100.00%	30.43%	0.00%	35.78%	0.00%	0.00%
West	0.00%	20.29%	100.00%	16.51%	0.00%	0.00%

Cluster 1

The mean age in this cluster is about 58. All individuals in this cluster are Whites from the southern part of the United States. All individuals in this cluster indicated that they would definitely vote in the 2012 Election.

Cluster 2

The mean age in this cluster is about 57. The majority of individuals are Whites from the various parts of the country (87 percent) and they indicated that they would probably vote in the 2012 Election (78 percent).

Cluster 3

The mean age in this cluster is about 57. All individuals in this cluster are Whites from the western part of the United States. All individuals in this cluster indicated that they would definitely vote in the 2012 Election.

Cluster 4

The mean age in this cluster is about 53. The majority of individuals are Blacks from the various parts of the country (64 percent). Nearly all individuals in this cluster indicated that they would definitely vote in the 2012 Election (98 percent).

Cluster 5

The mean age in this cluster is about 58. All individuals are Whites from the various north central parts of the country and they indicated that they would definitely vote in the 2012 Election.

Cluster 6

The mean age in this cluster is nearly 59. All individuals are Whites from the various northeastern parts of the country and they indicated that they would definitely vote in the 2012 Election.

Model Summary

Algorithm	TwoStep
Inputs	4
Clusters	6

Cluster Quality

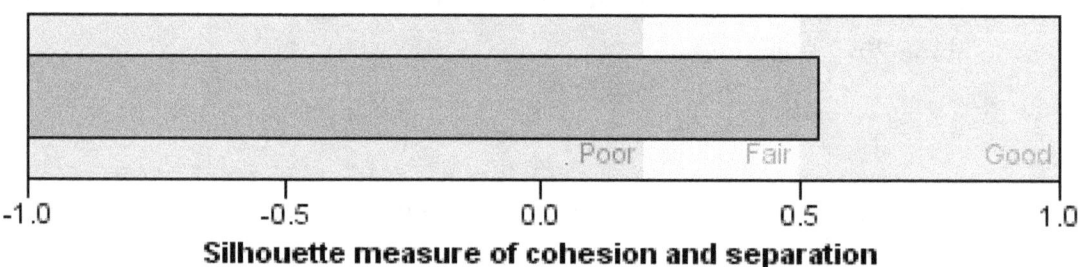

Silhouette measure of cohesion and separation

Example from 2007 Indonesian Family Life Survey (IFLS4)

```
TWOSTEP CLUSTER
 /CATEGORICAL VARIABLES = male MARSTAT Education district type
 /CONTINUOUS VARIABLES = AGE n_moves
 /DISTANCE LIKELIHOOD
 /NUMCLUSTERS AUTO 15 BIC
 /HANDLENOISE 0
 /MEMALLOCATE 64
 /CRITERIA  INITHRESHOLD (0)  MXBRANCH (8)  MXLEVEL (3)
 /PRINT COUNT SUMMARY
 /SAVE VARIABLE=TSC_9453.
```

Attribute	Cluster 1 (N = 400)	Cluster 2 (N = 327)	Cluster 3 (N = 680)	Cluster 4 (N = 622)	Cluster 5 (N = 517)	Cluster 6 (N = 560)
Age	34.93	34.25	35.65	31.72	22.18	29.45
Marital Status						
Not yet married	0.00%	0.00%	0.00%	0.00%	99.42%	0.00%
Married	100.00%	71.25%	100.00%	100.00%	0.00%	100.00%
Divorce / separated	0.00%	3.98%	0.00%	0.00%	0.00%	0.00%
Widowed	0.00%	11.31%	0.00%	0.00%	0.58%	0.00%
Sex (%)	0.00%	13.46%	0.00%	0.00%	0.00%	0.00%
Male	0.00%	4.59%	100.00%	0.00%	47.39%	0.00%
Female	100.00%	95.41%	0.00%	100.00%	52.61%	100.00%
Level of education						
Elementary	100.00%	16.82%	32.97%	39.07%	15.67%	0.00%
Junior high	0.00%	4.28%	20.15%	24.12%	18.76%	41.43%
Senior high	0.00%	9.48%	32.21%	36.82%	47.20%	58.57%
University	0.00%	69.42%	14.85%	0.00%	18.38%	0.00%
# Moves	0.38	1.34	0.81	0.72	0.58	0.71
Province where the facility is located						
Sumatra	0.00%	19.27%	22.79%	55.79%	17.99%	0.00%
Java	100.00%	62.08%	56.32%	0.00%	63.83%	100.00%
Other	0.00%	18.65%	20.88%	44.21%	18.18%	0.00%
Type of facility						
Public	31.50%	23.24%	24.26%	30.06%	29.98%	46.79%
Private	19.00%	42.81%	37.79%	20.74%	40.03%	23.93%
Other	49.50%	33.94%	37.94%	49.19%	29.98%	29.29%

Cluster 1

The mean age in this cluster is nearly 35. All individuals in this cluster are married females with at most an elementary school education and visited facilities located in Java. Individuals in this cluster are less mobile. The majority of individuals in this cluster use either public or other types (e.g. nurse, paramedic, midwife, and traditional practitioner) of outpatient care facilities. Less than one-fifth of the individuals in this cluster used the private facilities.

Cluster 2

The mean age in this cluster is about 34. The majority of individuals in this cluster are married (71 percent), females (95 percent), has a university education (nearly 70 percent) and visited facilities located in Java (62 percent). A somewhat higher percentage (43 percent) of individuals in this cluster used the private facilities. Individuals in this cluster are relatively more mobile than the other segments.

Cluster 3

The mean age in this cluster is between 35 and 36. All individuals in this cluster are married males with varying levels of education. More than half of individuals in this cluster visited facilities located in Java. The majority of the individuals in this cluster used the private and other types of outpatient facilities.

Cluster 4

The mean age in this segment is close to 32. All individuals in this cluster are married males with at most a senior high education. Over 50 percent of individuals in this cluster visited facilities located in Java. Another 44 percent of individuals in this cluster visited facilities located in other provinces like Sunda, Kalimantan, Sulawesi, Maluku, and Papua. Nearly half (49 percent) of individuals in this cluster used other types of facilities.

Cluster 5

Individuals in this cluster are relatively younger than the other clusters. The mean age in this cluster is about 22. The majority of individuals (99 percent) in this cluster are not married. Nearly half of the individuals in this segment have a senior high education and nearly 64 percent of the individuals in this segment visited facilities located in Java. A somewhat higher percentage (40 percent) of individuals in this segment used the private facilities.

Cluster 6

The mean age in this segment is 29. All individuals in this segment are married females with either a junior high or a senior high education. All in this segment visited facilities located in Java. A somewhat higher percentage (47 percent) of individuals in this segment used the private facilities.